PROCESS SAFETY AND BIG DATA

PROCESS SAFETY AND BIG DATA

SAGIT VALEEV
Department of Computer Science and Robotics,
Ufa State Aviation Technical University, Ufa, Russia
Faculty of Ecology and Engineering, Sochi State University,
Sochi, Russia

NATALYA KONDRATYEVA
Department of Computer Science and Robotics,
Ufa State Aviation Technical University, Ufa, Russia
Faculty of Ecology and Engineering, Sochi State University,
Sochi, Russia

ELSEVIER

Elsevier
Radarweg 29, PO Box 211, 1000 AE Amsterdam, Netherlands
The Boulevard, Langford Lane, Kidlington, Oxford OX5 1GB, United Kingdom
50 Hampshire Street, 5th Floor, Cambridge, MA 02139, United States

Library of Congress Cataloging-in-Publication Data
A catalog record for this book is available from the Library of Congress

British Library Cataloguing-in-Publication Data
A catalogue record for this book is available from the British Library

ISBN: 978-0-12-822066-5

Publisher: Susan Dennis
Acquisitions Editor: Anita A Koch
Editorial Project Manager: Ruby Smith
Production Project Manager: Kumar Anbazhagan
Cover Designer: Victoria Pearson

Typeset by SPi Global, India

Contents

Preface

Big data and process safety is a new interdisciplinary field that has formed at the intersection of information technology, management and control theory, and production safety management.

The successful application of big data technologies in the field of process safety is based on the knowledge and competencies of personnel in the field of large systems management and control, as well as hardware and software system developers and technologists.

About the book

Process Safety and Big Data discusses the principles of process safety and advanced information technologies. It explains how these are applied to the process industry and provides examples of applications in process safety control and decision support systems.

This book helps to address problems faced by researchers in industry that are the result of increased process complexity, and that have an impact on safety issues, which have also become more and more complex. The book presents ways to tackle these safety issues by implementing modern information technologies, such as big data analysis and artificial intelligence. It provides an integrated approach to modern information technologies used in the control and management of process safety in industry.

Book features

- Paves the way for the digital transformation of safety science and safety management
- Takes a system approach to advanced information technologies used in process safety
- Provides examples of how artificial intelligence is applied in the contextualization of the data streams that are monitored to improve safety performance
- Applies big data technologies to process safety

Audience

The book is primarily intended for researchers from academia and industry, who are working on process safety systems.

The book can also be used at the graduate or advanced undergraduate level as a textbook for courses such as process safety, control systems, computational science, and many others.

Organization of the book

This book consists of seven chapters.

1. Large-Scale Infrastructures and Process Safety
2. Risk and Process Safety Standards
3. Measurements, Sensors, and Large-Scale Infrastructures
4. Databases and Big Data Technologies
5. Simulation Technologies for Process Safety
6. Big Data Analytics and Process Safety
7. Risk Control and Process Safety Management Systems

Acknowledgments

As the authors, we hope that the book will stimulate further research in process safety provision, and apply the results of the research in real world applications. We hope sincerely that this book, covering so many different topics, will be very useful for all readers.

The authors are grateful to Anita Koch, Elsevier's Acquisition Editor, and the whole book's production team at Elsevier for their cooperation, help, patience, and courtesy.

Sagit Valeev
Natalya Kondratyeva

CHAPTER 1

Large-scale infrastructures and process safety

1.1 Introduction to system approach

To understand the problem of process safety management to minimize risks using big data technologies, we need to recall the concepts used in system analysis theory. The basic definition of system analysis is the definition of the concept of a system. There are many definitions, and they reflect the features of a particular problem solved in the framework of system analysis. In the context of our tasks, the elements and relationships between that form the system are important. Thus, a system is a group of interacting or interconnected objects that form a single whole (System, 2020).

Any system has its main goal. In our case, the process safety management system should provide a given level of safety. Entities (elements) of the system can have their own goals, used by the system to achieve its own goal.

The goals of the elements are formed based on the purpose of the system. To do this, a decomposition of the goals of the system is performed, and the elements and relationships between them to achieve them are determined.

The property of a system to form its goal based on the goals of the elements is determined and maintained within the framework of the concept of the life cycle of systems:

- At the stage of system design, the elements of the system are determined and the relationships between them are described.
- At the stage of functioning of the system, the operability of the elements is maintained and the quality of the connections between the elements is controlled.
- At the disposal stage, elements are identified that can be reused or recycled. If this is not possible, then the elements can be buried or destroyed in the external specialized system.

The system architecture on the designing stage can be presented in the form of subsystems and links between them.

The purpose of technological processes in our case is the conversion of raw materials into a final or intermediate product. That is why they can be

Process Safety and Big Data
https://doi.org/10.1016/B978-0-12-822066-5.00007-8

attributed to the elements of the production system of the final product. Modern production systems are defined by their spatial boundaries and technological processes by temporary boundaries. Any system function in another system surrounding it—for example, the environment—influences our production system and by itself is under production system influence.

In some cases a system can be described through the descriptions of its elements and the relationships between them, as well as through the descriptions of the functions it performs.

Note that the system has properties that are not inherent in its elements. This is the so-called synergistic effect.

When classifying systems, attention is paid to interaction with other systems. If there is an active exchange of raw materials, energy, or information with other systems, then they are classified as open systems.

An example of an open system is a modern petrochemical enterprise, in which oil products coming from outside are processed into the final product, and electricity coming from the power system is spent on technological processes. Closed systems do not exchange energy with the environment (another system).

When describing systems, a functional approach is used (describing the system as a black box), when only its input variables and output variables are specified for its description. The function of converting input variables to output data can be presented in the form of a description of its properties, without revealing the details of its implementation. This approach does not allow decomposing a complex system into subsystems, but can facilitate its analysis and modeling using various methods of the theory of identification of systems.

1.2 Large-scale infrastructures

Industrial infrastructures are located on land or in water, some of them can shift their location, and all of them have a number of typical features that characterize them as a special class of complex systems. Their features include the fact that they occupy a large area and volume, and their organizational structure includes many interconnected complex technical objects. The functioning of the large-scale infrastructure is supported by a large group of people organized hierarchically. All of them have different qualifications and professional knowledge.

Large industrial infrastructures are studied and researched in various fields of science and using different approaches:

- Systems approach. In this area, we look at systemic patterns that, among other things, affect process safety features (Leveson, 2016).
- Integrated adaptive systems approach. We are looking for solutions to the problems of adapting the control and management system taking into account changes in the system (Curry, Beaver, & Dagli, 2018).
- Cyber-physical systems approach. Here, when analyzing and studying the properties of a system, the features of information exchange between subsystems and the optimization of these processes based on computer networks are taken into account (Li, 2016).

Another feature of large infrastructures is that they use a large amount of electricity, or other types of energy sources. Therefore, there is an urgent requirement to optimize the processes of energy consumption and the use of alternative energy sources (Brennan, 2012).

One more feature of large-scale infrastructure is the application of modern computer systems for control of technological process and provision of efficient management procedures.

Depending on the purpose and need for access to the resources, infrastructures are located in hard-to-reach places, or, conversely, near megacities, if their operation requires the participation of a large number of people.

Infrastructure is often affected by various factors related to complex climate conditions. For example, oil platforms in the Arctic are exposed to low temperatures and gale-force winds. One of the main characteristics of industrial infrastructures is that they almost always pose a potential risk to the personnel working on them, as well as to the people who live near them.

Unfortunately, there are many examples of these hazards events (Sanders, 2015).

Depending on the characteristics of technological processes in industrial infrastructures, if the rules for performing technological processes are violated, as well as the rules for maintaining technological installations they can (and unfortunately do) cause significant damage to the environment (Nriagu, 2011).

Let's look at examples of industries where large infrastructure facilities are operated.

Petrochemical enterprises. Production and processing of oil and gas are associated with complex and expensive technological processes. Oil-producing land and sea systems are used to implement these processes. Processing of hydrocarbon raw materials is implemented at petrochemical enterprises.

Transportation of petroleum products is carried out using pipelines and various types of land transport and water transport.

Energy sector. Petrochemistry is closely related to the electric power industry, as technological processes require the use of large amounts of electricity. The objects of the energy industry include hydroelectric power stations, nuclear power plants, and reservoir systems. In turn, power plants use natural gas and other hydrocarbons to generate electricity.

Transport industry. This includes the aviation, railway, and automobile industries. These industries, in turn, include airports, train stations, bridges, roads, and river and sea ports. The transport industry is one of the main consumers of various types of fuel: kerosene, gasoline, fuel oil, and lubricants.

Industrial branch. Consumers of petrochemical products are various industrial enterprises that produce a variety of high-tech products, such as cars, tires, dishes, etc. Industry is also a consumer of large amounts of electricity. The data collection, storage, and transmission industry based on computer-aided information processing consumes significant energy resources. Data centers of large companies use electricity to cool their clusters, which generate heat during data processing.

We shall now look further at various infrastructure objects that are the parts of large industrial infrastructures.

1.2.1 Drilling rig

A drilling rig is a complex mechanical system that requires considerable effort from a team of specialists to install, maintain, and dismantle it. When performing technological processes, it is necessary to comply with safety rules, since the technical processes of drilling operations, installation of drilling column elements, and delivery of downhole equipment elements are associated with the movement of mechanical structures of large weight and include various manual operations. A modern drilling rig can be classified as a complex mechatronic system. It can contain a robotic system for changing and installing casing pipes, an automated winch, etc. To control the manipulator, a hybrid control system is used, including an automatic control system and control based on operator commands. The personnel serving the drilling rig may number more than 30. Since the drilling process is usually a continuous process, the work is performed in shifts. Putting a well into operation can take up to a year. This time depends largely on the type of well bore.

Fig. 1.1 Modern rig and main types of hazards.

To monitor the status of all subsystems of the drilling rig, the following data sources are used: pressure sensors, speed sensors, drill position sensors, temperature sensors, and various actuators—electric motors, motors, pneumatic automation systems, valves, etc. Fig. 1.1 shows a modern drilling rig and its simplified scheme and main types of hazards.

The most dangerous substances for team on the drilling rig include formation gas, hydrogen sulfide, and diesel fuel. Emergencies are related to electrical equipment, cable breakage, and gas release from the well, and destruction of the unit's structures due to corrosion and fatigue stresses.

1.2.2 Oil offshore platform

An offshore oil platform for the extraction or processing of hydrocarbons operates under extreme climatic conditions and the impact of dynamic loads on its structure. This is due to high humidity, exposure to solar radiation and strong wind, sea water, and waves. In addition to the usual subsystems for an oil rig, it includes subsystems for ensuring the safety of infrastructure from the influence of sea conditions, communications, and rescue vehicles (Khan & Abbassi, 2018). Fig. 1.2 shows an offshore oil production platform that includes a platform, logistics support systems, a drilling rig, and a life support system.

The operation of the oil platform is associated with various hazards, which may cause emergency situations, with loss of performance and with threat to the life of personnel. The main types of hazards are presented in Fig. 1.2.

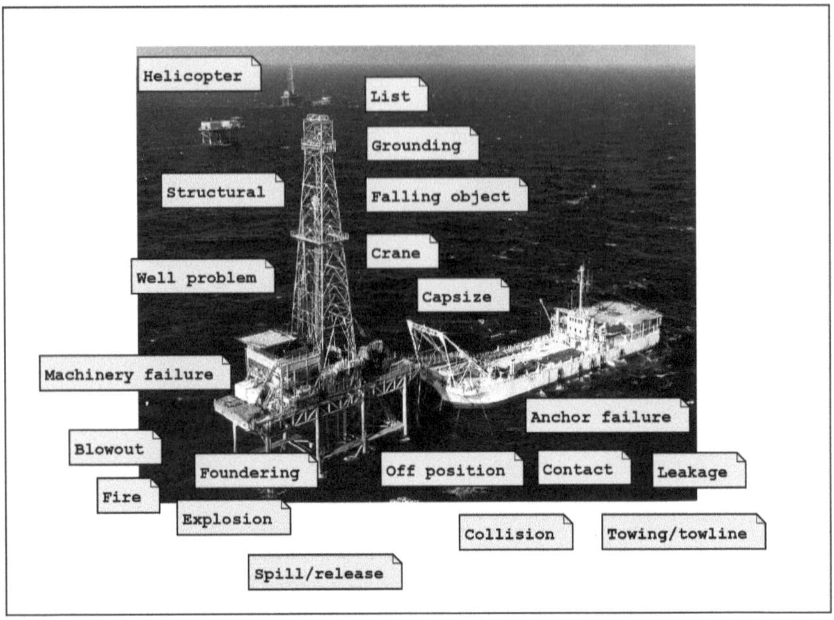

Fig. 1.2 Offshore oil rig and events hazard classification.

1.2.3 Tanker for transportation of liquefied gas

Most of the oil and gas fields are remote from the places of their processing and consumption of marketable oil and gas products. Oil prepared at the oil field is transported to the refinery by pipeline, rail, or water (sea and river). The main types of hazards for a pipeline are caused by damage to metal construction through the corrosion processes.

Liquefied gas is transported on a specialized tanker. The tanker includes the following main subsystems: liquefied gas storage systems, the vessels for transporting gas storage facilities, and systems for maintaining optimal characteristics of liquefied gas. Transportation of liquefied gas is relatively dangerous (Mokhatab, Mak, Valappil, & Wood, 2014). Fig. 1.3 shows a tanker transporting liquefied gas and events that can lead to hazardous situations.

When transporting various types of fuel by land vehicles, specially equipped railway tanks or car transport systems are used (see Fig. 1.4). The transportation of these products requires enhanced security measures and is associated with high risks of critical situations.

To maintain the working pressure in the pipeline, gas pumping systems are used, and may be equipped with gas turbine units. On main gas pipelines, compressor stations are equipped with centrifugal compressors with a gas

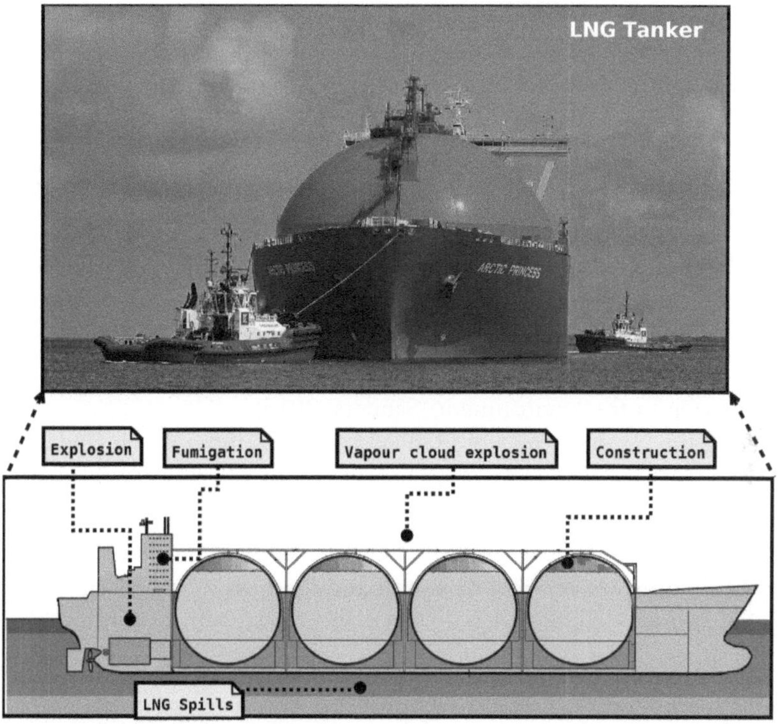

Fig. 1.3 Transportation of liquefied natural gas.

Fig. 1.4 Trucks are used for the transportation of diesel and petrol/gas/gasoline/benzin.

turbine drive or an electric drive. Here hazards include fires and explosions (Mokhatab, Poe, & Mak, 2015).

Gas pipelines can also be laid along the bottom of the sea; the pipeline copies the profile of the seabed, bending under its own weight. In order to ensure reliable operation of a modern gas pipeline, to monitor the pressure in the underwater part of the gas pipeline and the gas velocity in it, and to localize an emergency gas leak quickly, parallel installation of fiber optic cable is possible.

To control the state of the pipeline, monitoring systems are used, which include a large number of different devices for measuring the state of pipeline elements. The main hazards include explosions, fires, and leaks of toxic gases.

1.2.4 Petrochemical production

Oil and gas processing are performed at a petrochemical plant. The petrochemical plant contains a sufficiently large number of oil processing units connected by a single technological cycle. The main hazards are fires, explosions, leaks of toxic materials, the complexity of personnel evacuation, and large damage to the environment (Sanders, 2015).

Petroleum products are stored in oil storage facilities that include transport infrastructure, a pipeline system, many land tankers, and fire safety systems. The main hazards to personnel and residents of cities are leaks of toxic substances and oil products entering the environment. Fig. 1.5 shows a diagram of the primary oil treatment unit and the main types of hazards.

The various pumping systems, shut-off valves, and process monitoring devices are used for pumping petrochemical products. The petrochemical company can include power supply systems and by-product recycling systems (Mannan, 2013).

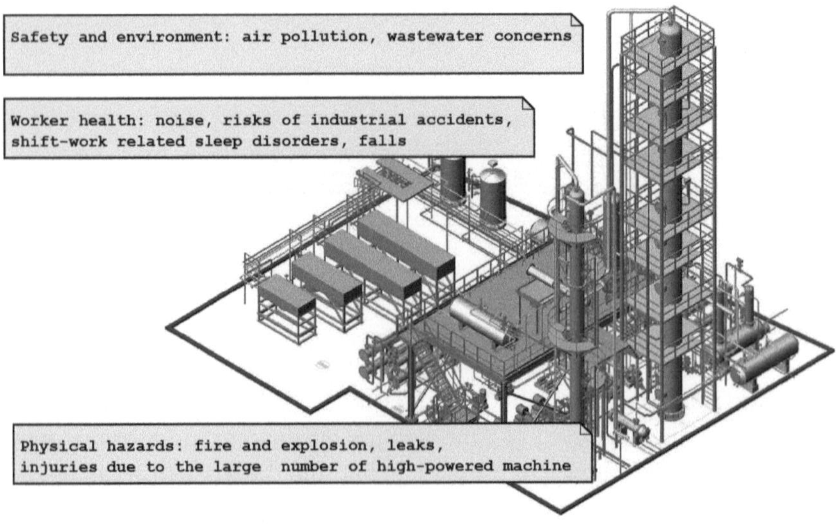

Fig. 1.5 Primary oil refining unit.

1.2.5 Industrial enterprise

Petrochemical products such as gas, gasoline, plastics, etc. are used in various industries: metallurgy, transport systems, and power generation at power plants. The main types of hazards are fires, explosions, leaks of toxic substances, damage to the environment, and the difficulty of evacuating personnel.

In modern industrial enterprises, various products are obtained from the products of oil refineries. Technological processes are based on application of different industrial units. They can include gas flaring equipment, a waste storage system, and a production waste disposal system. The main types of hazards to personnel and the environment are related to the toxicity of applied materials and possible emissions of hazardous substances during their disposal.

1.2.6 Monitoring and data collection systems

Monitoring and data collection systems allow us to measure the state of the characteristics of technological processes. The features of industrial measuring systems are high requirements for their accuracy and reliability. To collect and transmit data, controllers, computers, and local area networks of the enterprise are used.

For example, fire safety systems are automated systems and include smoke detectors, heat detectors, and fire detectors. In the event of a source of ignition using sensors, it is possible to localize the source and send an alarm to the control center. In fire safety systems, it is possible to suppress the source of ignition using fire extinguishing means. Fire safety means include sprinklers, automatic ventilation systems, automatic systems for blocking access to the room, and systems of stop valves. All these elements are used to control fire pumps and to supply substances for extinguishing fires (Fig. 1.6).

Industrial plants monitor the environment. Special equipment estimates air quality and the level of emissions of harmful substances. To monitor the state of the environment (for example, to analyze water quality), distributed measuring systems based on modern telecommunication systems are used. These systems can measure the state of water in a pipeline by processing and transmitting data on its quality. Solar panels are used as sources of electricity for the measurement system.

Industrial controllers are used for technological processes, managed by supervisory control and data acquisition systems (SCADAs). Information

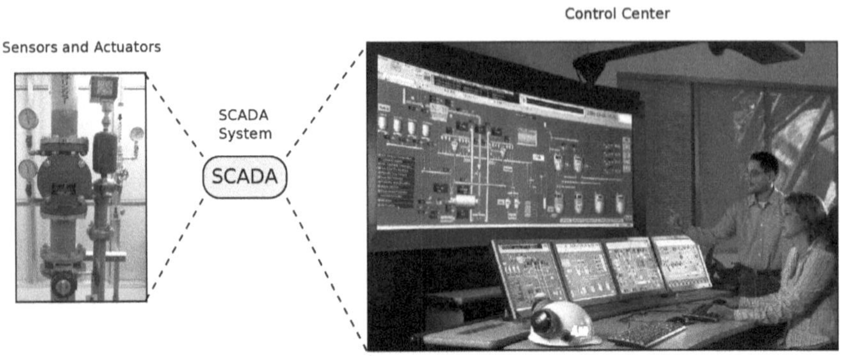

Fig. 1.6 Fire control system and SCADA control system.

in real time on the status of actuators and the state of technological processes is displayed on the screens of operators' monitors in the control center. The personnel responsible for the quality of technological processes receive up-to-date information and, taking into account the received data, can influence the course of technological processes.

1.2.7 Data center

Data collected at various stages of the life cycle of a large-scale infrastructure is stored in data warehouse systems based on data centers (Wu & Buyya, 2015). A modern data center occupies a large area and includes subsystems for uninterrupted power supply to hundreds of thousands of servers, routers, cooling systems and telecommunication systems (see Fig. 1.7). The data center itself is thus a large infrastructure object.

Data processing is carried out on specialized computers that are assembled in cluster systems. Fig. 1.7 shows server racks with clusters. For continuous calculations, clusters must be provided with uninterrupted power supplies. For reliable storage of data arrays, they are duplicated and stored on other servers (Wu & Buyya, 2015).

Data centers typically contain hundreds of thousands of computers, which consume large amounts of electricity and generate significant amounts of heat.

The main dangers for data centers are power outages, cluster overheating, data loss, and fires. Attacks on data center assets by hackers can cause significant harm. Therefore, special attention is paid to information security issues.

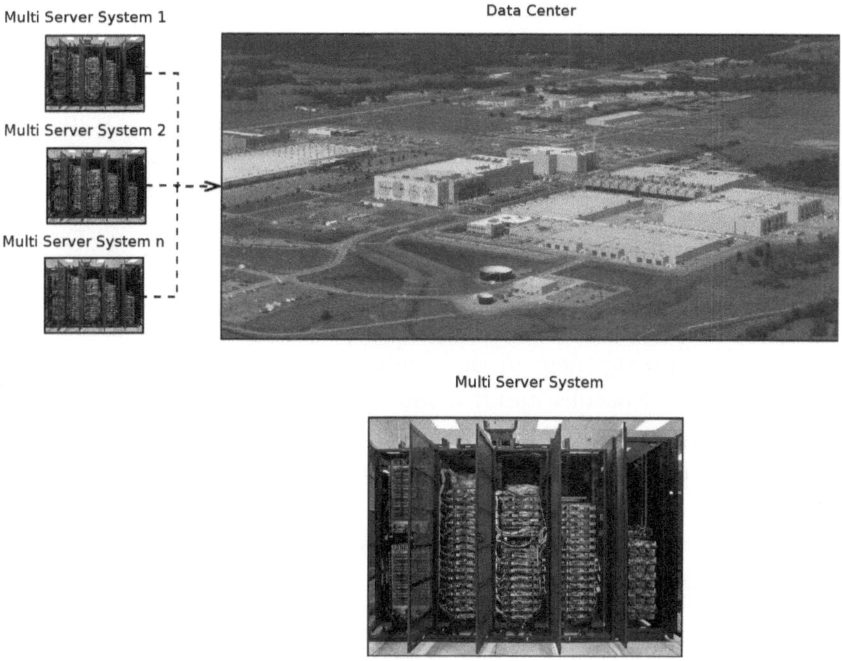

Fig. 1.7 Multiserver systems and data center.

It should be noted that modern large infrastructures are equipped with their own subsystems for collecting, storing, and transmitting data. Various features are inherent in subsystems for collecting and processing industrial data as follows:

- The problems associated with the collection of current data on the state of technological processes. This is due to the special conditions for the implementation of a number of technological processes: high temperatures, high pressure, aggressive environment, etc. (*Process Safety Calculations*, 2018).
- There are problems of data transmission from various subsystems under the influence of electromagnetic interference on radio channels, the difficulty of providing a continuous communication field (Tekin, Pleros, Pitwon, & Hakansson, 2017).
- Data storage requires solving many specific problems related to the processes of data collection, storage, and aging (Wu & Buyya, 2015).
- There are problems with processing large amounts of data (Wu & Buyya, 2015).

At present, the use of digital twins of complex technical objects for ensuring process safety, including security issues, is being actively discussed. When solving the problems of developing and building digital twins of infrastructure and infrastructure objects, many problems arise, such as the creation of digital models taking into account the characteristics of each instance of the class of infrastructure objects.

1.3 Problems of safety for large-scale infrastructures

Process safety management in large industrial infrastructures is associated with a number of peculiarities that should be mentioned. We shall take a closer look at them here.

There are no precise instructions for performing the sequence of steps to minimize risks in a critical situation. A good example is the approved evacuation plan in the case of an emergency. This plan usually contains a sequence of key actions to minimize risks. The generalized steps that must be performed if necessary are presented in standards and procedures for the prevention of damage in the event of a critical situation. However, it is difficult to predict in what exact place a critical situation will arise. Our emergency plan is therefore only good in common cases.

It should be noted that process safety is associated with the processing of statistical data used in calculating risks and in calculating various scenarios for responding to a critical situation (Johnson, 1973).

If we are able to make a plan based on statistical information, the most probable hazard location can be predicted. In this case, it is possible to focus on the main measures to prevent a critical situation in this area.

Thus, we need the development and application of advanced technologies for collection, transmission, storage, and processing of large amounts of data. This allows us to solve problems associated with identifying the characteristic paths for the development of critical situations in oil and gas, petrochemicals, and transportation industries.

In the ISO31010:2019 standard, considerable attention is paid to risk assessment methods, taking into account the amount of necessary information (IEC 31010:2019 Risk management—Risk assessment techniques, 2019).

These data sets can also be used to clarify the standards for process safety, more effective training procedures for personnel, and the development of various models for analyzing critical situations.

One of the features that we consider when solving the problems of minimizing risks in industrial infrastructures is that they are subject to the influence of various factors of uncertainty associated with internal and external influences.

These factors include, for example, inconsistent changes in technological documentation and violation of safety regulations. Possible errors may have been made at the preliminary design stage—for example, designers didn't take into account during the design process that electric units would be operating in seismically dangerous zones (Health risk assessment from the nuclear accident after the 2011 Great East Japan Earthquake and Tsunami, 2013).

Thus, the properties of a complex infrastructure can change over time and therefore the possible places of emergence of a critical situation can be changed.

When developing models of a critical situation, the data obtained during the measurement of technological process parameters are used, and modeling the results can also serve as a source of important information. From the point of view of catastrophe theory, a slight change in the parameters can lead to an abrupt change in the properties of a complex system. Unfortunately, it is not always possible to track the accumulation of changes in the organizational structure.

Large infrastructures can be classified as socio-technical systems. In these systems, the hierarchical management and decision support process depends heavily on personnel qualification. Thus, the safety of processes depends on both the management system and the level of staff training.

Due to changes of the infrastructure over time, it is difficult to assess the presence of a weak link in the safety system.

Effective implementation of big data technologies to ensure process safety requires additional costs, identification of stakeholders, training of personnel and development of standards for using big data to solve the problems of minimizing risks.

1.4 Complexity and large-scale infrastructures

Large infrastructures and their objects seem intuitively difficult to describe, and predicting their state is considered a complex and not always solvable task. As noted earlier, infrastructures can be dangerous to people and the environment. To find ways to reduce risks in emergencies, we need to understand the complex nature of infrastructures. This is necessary in order to understand how to solve the problem of monitoring the state of infrastructure, as well as to understand the price that must be paid to ensure its safety.

Suppose that events related to infrastructure can be divided into those that allow achieving the goal of infrastructure, and those that, on the contrary, do not allow achieving it. Moreover, we can definitely determine to which class the processes belong: constructive or destructive. Unfortunately, it is not always possible to solve this classification problem, due to its complexity.

The key concepts used to analyze the behavior of complex technical systems are *the state of the system* and *the process.*

If the state of the system is represented as a point in some n-dimensional space, then the motion characteristics of this point in this space characterize the properties of the process. That is, a process is a change in the coordinates of a point in the state space of a system that can be described or predicted. If we cannot predict the behavior of the process, then one of the solutions is to build a static process model. In some cases, it is possible to construct a dynamic process model based on a system of differential equations.

The simplest state space can be represented as a system of Cartesian coordinates of two or three variables. An example of such a space in the simplest case is a change in the process parameters depending on two parameters—for example, temperature and pressure. The state of the system is determined by its coordinates in the state space. These coordinates represent the values of the parameters of technological processes and parameters of the infrastructure itself. The state of the technological process at any moment of time can be described by the behavior of a point in the state space of the technological process. By process, we mean the transition of the system under consideration from one state to another state. In our case, we consider technological processes associated with the production, transportation, storage, and processing of gas, oil, and petrochemical products.

The process safety system is a management system that solves the problem of minimizing risks in the presence of potential threats arising from the influence of internal and external factors associated with the transition of the system from state to state during various technological operations.

The purpose of the process safety system is to reduce the risk to a level that is practically possible to implement using measures, rules, and technical means to prevent the occurrence of fires and explosions, as well as accidental or unforeseen spills, releasing chemicals that can harm human life and the environment. This system includes various subsystems: preventing leaks, oil spills, equipment breakdowns, monitoring over and under pressure, overheating, corrosion, and fatigue of metal structures. The system under consideration has a long life cycle, including the stages of detailed design, operation of facilities, maintenance, and disposal. At all stages, a large amount of heterogeneous information is generated, which includes arrays of data, entries in inspection logs, and audio and video files.

As we noted earlier, the state of a complex technical object at the current time is determined by the values of the set of parameters. It is assumed that the values of these parameters can be measured or calculated using digital models or digital twins.

Consider the example of the movement of a point characterizing the state of the technological process for three different situations, including:

- a situation of normal (regular) process execution;
- a situation of failure with the possibility of continuing after the restoration of the functioning of the system; and
- a critical situation where the process cannot be completed.

Let us imagine the trajectory of a point characterizing the state of the technological process in the form of a set of points $S(i)$, where $i = 0 \ldots n$. The process begins at the point $S(0)$ and should end at the point $S(n)$.The location of these points, for clarity, is given in three-dimensional space and is determined by three parameters, P_1, P_k, and P_s (see Fig. 1.8).

The transition of a technological process from one state to another is carried out according to the rules established by the technologist on the basis of the regulations for the implementation of the technological process. The deviation of the values of the process parameters at a given point can vary in an acceptable range of values, for example, for $S(0)$ this is the region $S\_d(0)$. The values characterizing this area are determined taking into account the requirements for the accuracy of maintaining process parameters when designing a technological system.

If the process parameters at the point of the operating mode go beyond the boundaries of the permissible region, as in the case shown in Fig. 1.8 from the point of the state space $S(m)$ to the point of the state space $S\_f(m)$, which fell into the failure zone Z_f, then this can lead to a high probability of malfunction and disruption of the process flow.

If during the restoration work it is possible to eliminate the malfunction, then the process can be continued. In Fig. 1.8, this is the transition from the state $S\_f(m)$ to the state $S(m)$.

If going beyond the permissible zone of changing the parameters of the technological process leads to the failure of the technological equipment, or

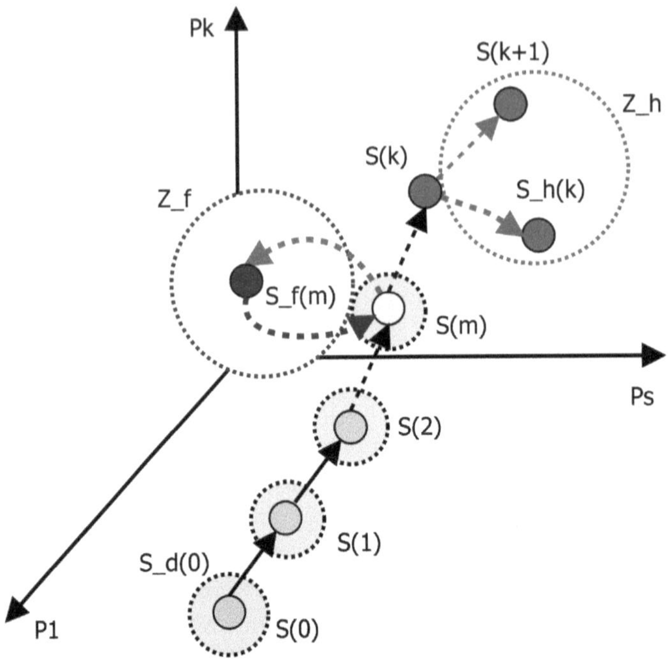

Fig. 1.8 Space of process parameters and possible trajectory of process state.

deviation of product quality from the required ones (defective situation), then the technological process cannot be resumed.

So, for example, in Fig. 1.8, the transition from state $S(k)$ to state $S\_h(k)$ of the critical situation zone Z_h does not allow the technological process to go to the next allowed state $S(k+1)$.

It is assumed that the controlled technological system is identifiable and observable—that is, the values of its coordinates can be measured, or determined with a given accuracy using sensors, or calculated using simulation models. These models should provide the required accuracy in determining the values of system parameters at a given point in time. Moreover, the transition from state to state in the state space can be continuous or discrete, as in our example. When solving the problems of ensuring the safety of technological processes, both options are possible.

The complexity of solving the problem of managing a technical system is determined by a number of factors as follows:

1. To determine the coordinates of the system in the state space, it is necessary to measure and calculate the values of the observed technological processes. This is not always possible to do for many reasons: there is no

appropriate measurement system, quality indicators are measured on the basis of laboratory tests, and not in real time, etc.

2. The transition from one state to another should be monitored taking into account changes in the parameters of technological equipment due to depreciation, or violation of operating rules, as well as deviations in the quality of raw materials from specified requirements.
3. It is necessary to ensure control and prevent the process parameters from entering the fault zone and the critical situations zone. This is one of the most important tasks of ensuring the safety of technological processes.
4. Errors of operators and personnel during the operation of equipment and improper fulfillment of the requirements of the regulations for the performance of the process.

In the state space, in which the state of the technological process is reflected, many processes can coexist at the same time, and they can exchange energy or information with each other.

To manage and control all related processes to achieve the main goal, it is necessary to solve the problem of multivariable process control task—that is, control of many interconnected processes simultaneously. If the system is manageable, then this problem can be solved; if the control actions are less than the controlled parameters of technological processes, then solving the problem requires the development of a control system taking this fact into account.

The question arises of how complex it is to implement management processes to achieve the global goal of the system. To assess the complexity of a technical system, various metrics or complexity scales are used. Typically, when analyzing the complexity of a system, the number of elements in the system and the relationships between elements are taken into account.

Consider a system consisting of two interconnected elements $s(1)$ and $s(2)$. Elements can exchange information or energy, and they can also change their state. Fig. 1.9 shows a directed graph that describes our simple system,

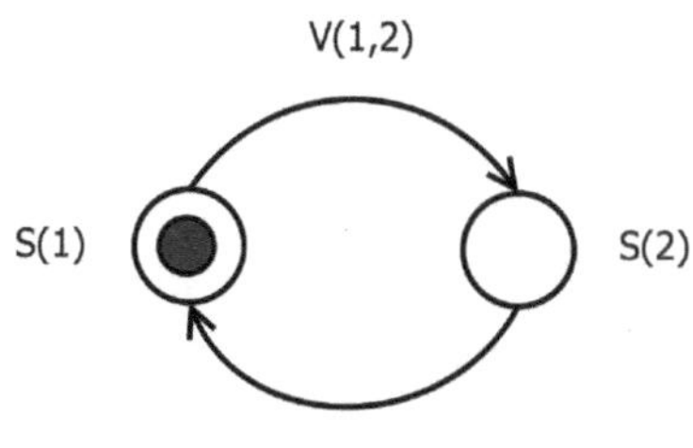

Fig. 1.9 System *S* consisting of two elements.

where $S = (s(1), s(2))$ are the nodes of the graph or elements of the system, and $V = (v(1,2), v(2,1))$ is the relationship between the elements that determines the direction of the exchange of information or energy. The exchange of information or energy allows elements to transition from one state to another. The transition from state to state may occur randomly or at the command of a control system.

Elements of system S can be in one of two states: "0" or "1." Therefore, the system can have one of four states: (0,0), (0,1), (1,1), (1,0), where the element in state "1" in Fig. 1.9 is identified with a marker.

The task of analyzing system S is to determine the states of its elements. If we know the sequence of commands and we can determine the state of an element using the means available to us, then the problem is solved. In the case of random transitions from state to state, to predict the behavior of the system S, information is needed on the history of its states.

We need an assessment of the complexity of the managed system to assess the complexity of solving the management problem. Next, we consider the task of assessing the complexity of a large-scale system.

The complexity of the system can be represented as a function $C = f(t, n, V, p(n), q(V))$, where t is time, n is the number of elements, V is the number of connections between elements, p is the procedure for determining the state of system elements, and q is the procedure for determining the state of elements V.

If the studied system includes a large number of elements, and they can be in different states, then the complexity of analyzing the state of the system increases significantly.

Let us estimate the number of procedures required to analyze the state of a large system, the model of which is a fully connected graph—that is, it is assumed that all elements of the system are interconnected.

As can be seen from Table 1.1, if there are more elements in a large system, and if they are all interconnected, then the number of possible connections is very large. The number of procedures for analyzing the state of the elements of a large system linearly depends on the number of elements in the system. The number of procedures for analyzing the relationships between elements, in our case, is $n(n-1)/2$, meaning that the algorithmic complexity is $O(n^2)$.

Analysis of the state of the elements of the studied system usually requires certain material and time costs. If system elements can take different states, then different analysis procedures may be needed to analyze the states of

Table 1.1 The number of procedures necessary for the analysis of the system status.

n	*V*	*p*(*n*)	*q*(*n*)
2	2	$p(s(1)) + p(s(2))$	$q(s(1)) + q(s(2))$
3	6	$p(s(1)) + p(s(2)) + p(s(3))$	$q(s(1)) + \cdots + q(s(6))$
10	45	$p(s(1)) + p(s(2)) + \cdots + p(s(10))$	$q(s(1)) + \cdots + q(s(45))$
100	4950	$p(s(1)) + p(s(2)) + \cdots + p(s(100))$	$q(s(1)) + \cdots + q(s(4950))$
1000	499,500	$p(s(1)) + p(s(2)) + \cdots + p(s(100))$	$q(s(1)) + \cdots + q(s(499{,}500))$
1,000,000	499,999, 500,000	$p(s(1)) + p(s(2)) + \cdots + p(s(1{,}000{,}000))$	$q(s(1)) + \cdots + q(s(499{,}999{,}500{,}000))$

these elements. It should be noted that the states of the elements and their characteristics usually depend on time and on the state of other elements of the system, which also affects the complexity of the analysis of large systems.

Industrial systems are open systems because other systems interacting with them influence their functioning. Given that industrial facilities and their control systems can be attributed to the class of nonequilibrium systems, their state can change significantly with a slight change in the parameters of some of their characteristics.

We shall now consider a few examples of various technological systems associated with oil production and refining, paying particular attention to the complexity of these systems.

1.5 Graph model of large-scale systems

Metal structures are the skeleton of all technological objects used in oil and gas production and their processing in petrochemicals. These are complex structures that allow you various items of equipment to be placed and secured: electric motors, pumps, power systems, winches, drilling tools, valves, etc.

Elements of metal structures are connected by welds, bolts, and rivets. The quality requirements for the steel used in the construction are very high. Metal structures are checked during manufacture for defects. Welds are subjected to strict control by various methods of technical diagnostics: external inspection, inspection of welded joints and structural elements using nondestructive testing devices.

To assess the complexity of the technical system, we use the methods of system analysis. We perform the decomposition of the system into subsystems, identify the main elements, and determine the relationships between them. After that, the structure of the system can be represented as an undirected graph. Analysis of the graph allows us to evaluate the complexity of the system and determine the elements and relationships that affect to a greater extent the structural integrity of the system.

1.5.1 Oil rig system graph model

A modern oil rig is a complex mechatronics system (Onwubolu, 2005). It includes a module for lifting drill string elements, a module for a drilling tool, a base, a tower, an equipment control system, and a data collection system on the status of technological processes. Therefore, the drilling rig can be assigned to the class of a complex technical system, including many elements and their relationships.

Note that the model reflects the structural properties of the system. This model does not take into account the properties of the system associated with fires and explosions. To analyze these factors, it is necessary to develop another dynamic model based on the scenario approach. In the process of assembling complex technological equipment, the modular principle is used, whereby complex structures are assembled from modules, which in turn consist of other modules.

During assembly, the modules are attached to each other using welds or other methods. As a result, the assembly unit is a complex technical object with new properties that differ from the properties of the modules. To analyze the state of structural elements, data obtained during the external examination and the readings of various sensors are used. This information is transmitted to information systems and stored in databases.

Fig. 1.10 shows a simplified diagram of a drilling rig, and also presents its model in the form of an undirected graph. The graph includes the following vertices: A is tool (drill bit), B is clamps, C is heavy drill pipe, D is drill pipes, E is casing pipes, F is well, G is DC motor, H is transmission, I is clutch, J is brake, K is winch, L is steel cable, M is tower, N is pulley system, O is top drive drilling engine, P is dead end cable support, Q is hook load sensor, and S is basic design. Next, we consider the analysis of the oil derrick as a system.

The model of the system, which describes the structural relationships of the elements of the oil rig, includes several interconnected subsystems: S_1 is electric traction drive (lift), S_2 is drilling and well drilling system, S_3 is construction, and T is the team of personnel. Preserving the health of personnel

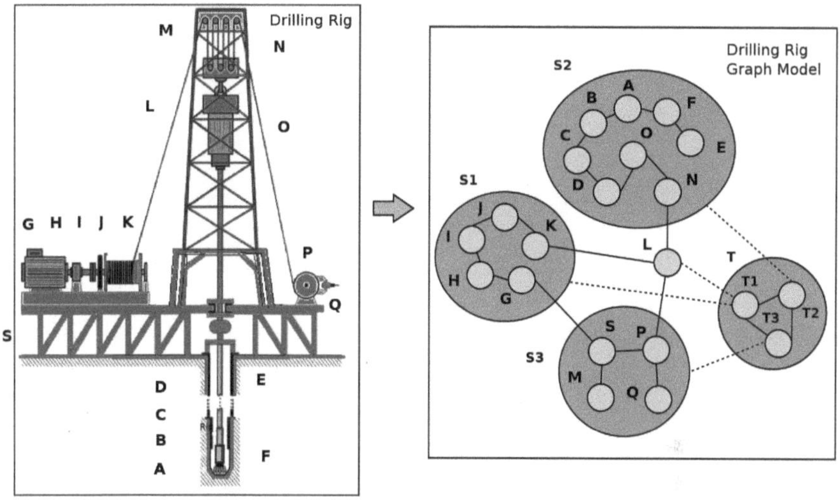

Fig. 1.10 Rig and graph model.

is the main task of ensuring the safety of technological processes. On the other hand, the prevention of hazards depends on the actions of personnel. Monitoring the condition of the equipment is assigned to a competent member of the team, designated as T_1.

The system graph includes 21 vertices and 25 basic relationships between the vertices. Analysis of relationships and elements allows us to identify critical relationships and elements of the system. In our case, the critical element is the L node and the L-T_1 link.

The main element for our case is the steel cable L. It connects all the subsystems of the rig. In the event of a break, the system breaks up into disparate subsystems and cannot ensure the execution of work, meaning that it cannot ensure the achievement of the main goal. A member (T_1) of the crew servicing the rig conducts a regular inspection of the state of the L cable. The system's performance depends on this person.

As a result of the analysis, it was possible to identify weak links in the system. Although it is clear what is needed to solve the safety problem of technological processes, solving this problem is still a complex procedure.

1.5.2 Oil platform system graph model

Using the simplified model of an oil platform as an example, let us consider in general terms the process of forming a complex system consisting of modules, as well as the process of collecting data on the state of system elements (Khan & Abbassi, 2018).

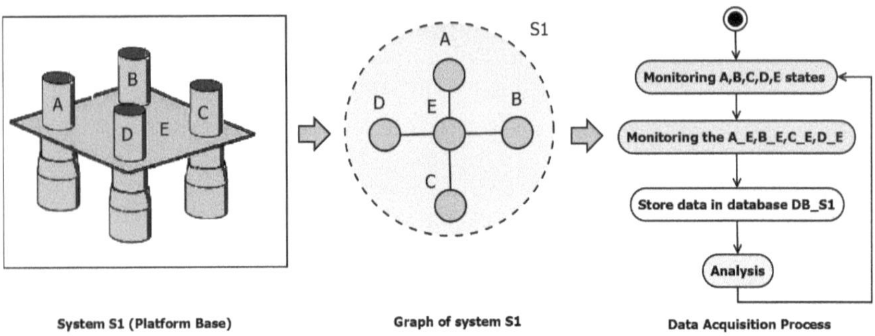

Fig. 1.11 Platform base.

Fig. 1.11 shows the unit that forms the base part of the oil platform. The structure is depicted in the form of a graph $S_1 = ((A,B,C,D,E),\{L_1\})$. The main element is the element E, which connects the columns of the platform. The condition of the structure is determined by periodic inspection of the columns by specialists, and the results of the inspection are recorded in the logs and transmitted as data to the DB_S_1 database. The set of links $\{L_1\}$ defines the connections of all elements with each other, thereby determining the complexity of the system S_1. Data on the results of monitoring the state of the elements are analyzed, and the findings are also transferred to the database.

When installing the rig on the platform, a new system is formed, including the base of the platform S_1 and the oil rig, presented in the form of system $S_2 = ((A_1, B_1, C_1, D_1)),\{L_2\})$ (Fig.1.12). The L_2 connection between the S_1 and S_2 systems is ensured by a reliable fastening system. The state of the

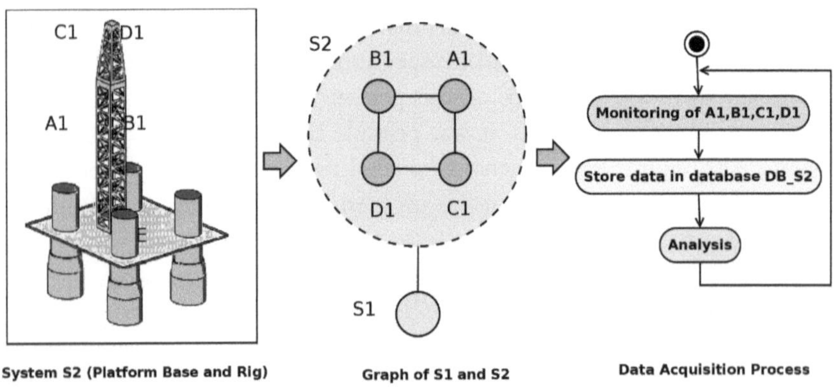

Fig. 1.12 Platform base and drill rig.

structure and elements of the drilling rig is determined by periodic inspection, and the state of the elements of the equipment is monitored by measurements obtained by sensors and periodic results of the analysis of the structural elements. These data are transferred to the company database DB_S_2.

The oil platform includes two decks on which power plant, fuel supplies, drilling elements, life support systems of the drilling crew, cranes, helipad, and communication and navigation systems are located. As a result, the complexity of the object grows as new subsystems $S_3 = ((A_{2\_1}, A_{2\_2}, A_{2\_3}, C_2, D_2), \{L_3\})$ (see Fig. 1.13) and $S_4 = ((A_{3\_1}, A_{3\_2}, A_{3\_3}, A_{3\_4}, C_3, D_3), \{L_4\})$ (see Fig. 1.14) appear.

It should be noted that in some cases, the platform may carry out oil refining for the internal needs of the platform. Thus, the risks associated with possible fires and explosions increase.

Fig. 1.14 shows an element of the system: the landing pad for the helicopter D_3. During takeoff and landing, the air currents generated by the helicopter propeller blades can move objects that are not fixed from the platform surface, which can cause personal injuries. Information about the state of elements of systems S_3 and S_4 is stored in the databases DB_S_3 and DB_S_4.

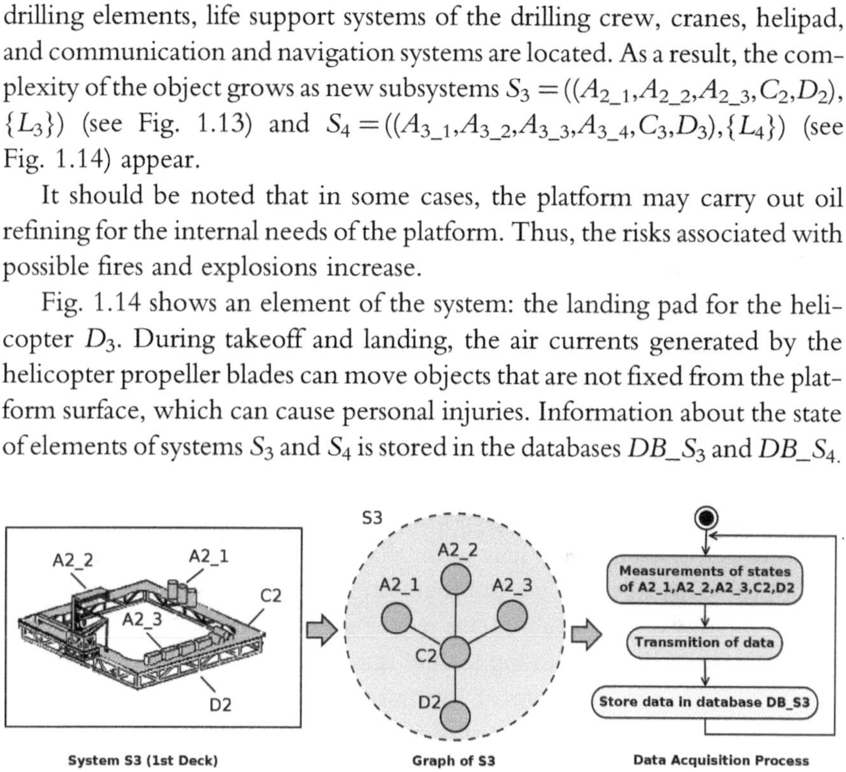

Fig. 1.13 The first deck of the platform.

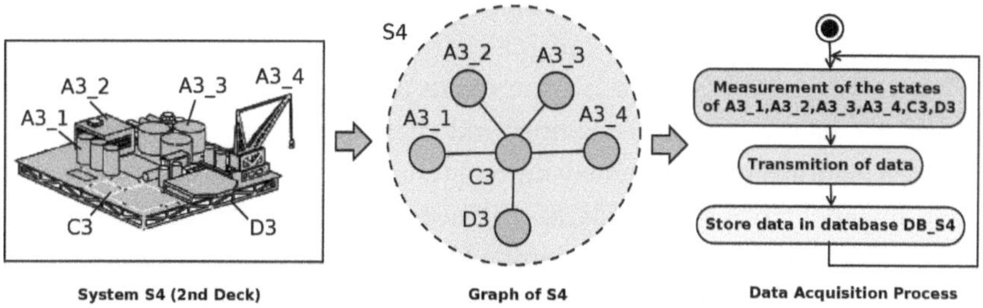

Fig. 1.14 The second deck of the platform.

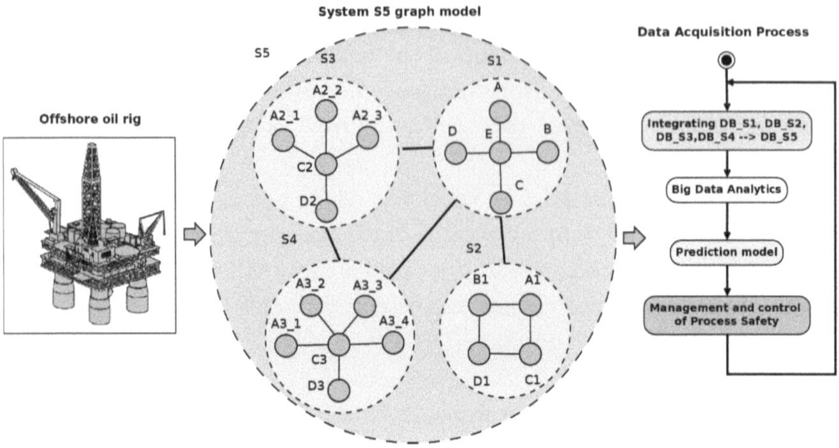

Fig. 1.15 Oil platform, graph model, data acquisition process.

In the process of forming the graph of the "oil platform" system $S_5=((S_1,S_2,S_3,S_4),\{L_5\})$ (Fig. 1.15), its complexity increases. Moreover, when forming the graph, only some subsystems and their relationships were used. The considered graph reflects only the structural relationships between the subsystems. Data on the state of the platform subsystems are periodically sent to the databases, and we can then monitor the status of the system, as well as the impact of its subsystems on the environment.

A feature of data processing for the system in question is the need for:

- solutions to the problem of integrating multiple databases that store information about the state of structures, technological processes, and the state of the environment;
- various analytical algorithms for heterogeneous data of large volumes; and
- a solution to the problem of building a predictive model of the state of platform elements and a probabilistic risk model.

When solving all these problems, the results of the analysis of a large amount of data can be used to build a control and safety control system. Creating an offshore platform is an extremely difficult engineering task, requiring significant material costs, the skills of engineers and workers, as well as the patience of the entire team working on the project.

From the point of view of ensuring the safety of technological processes, it should be borne in mind that the elements of the platform are subject to aggressive environmental influences, and the platform personnel work in difficult conditions, so the efficiency of the platform depends on the fulfillment of all safety rules.

1.5.3 Oil refinery system graph model

One of the most complex and expensive infrastructures we are considering is an oil refinery. It includes many different production units for oil refining and separation of its fractions, connected by a network of pipelines. Technological units are made of special steel, providing corrosion resistance. In the implementation of all technological processes, operations associated with high temperature and high pressure are used. For the pumping of intermediate products, pumps of various capacities are used, fed via electric networks. A large set of valves allows the flow of petroleum products to be controlled.

It should be noted that the required level of technological process safety is provided at various system levels of petrochemical enterprise management. We should take into account the great complexity of the infrastructure, which functions under the influence of various uncertainty factors. These factors include: fatigue stresses of structures, planned wear of equipment, possible violations of the tightness of various valves, violations of safety rules by personnel, etc.

An oil refinery is a connecting link that unites various transport industries: sea, aviation, and land transportation. The plant's products are used in the production of various materials used by other industries. Fig. 1.16 shows the model of the oil refinery in the form of a graph. The input source to the plant is crude oil, and the outputs are diesel, aviation fuel, marine fuel, gasoline, and other intermediate products.

All technological units are connected by a network of pipelines providing continuous communication between technological processes. This model allows us to analyze the movement of material flows and build a model for the distribution of risks associated with the processing and transportation of petroleum products.

When comparing the complexity of objects, a scale is chosen on which the difference between the object or system complexity from each other is determined relative to this scale. We chose the number of elements and relationships between them as a scale, which allows us to evaluate the complexity of the system from the point of view of analyzing the state of the elements and the relationships between them, to determine the movement of material flows and information flows, to build risk models.

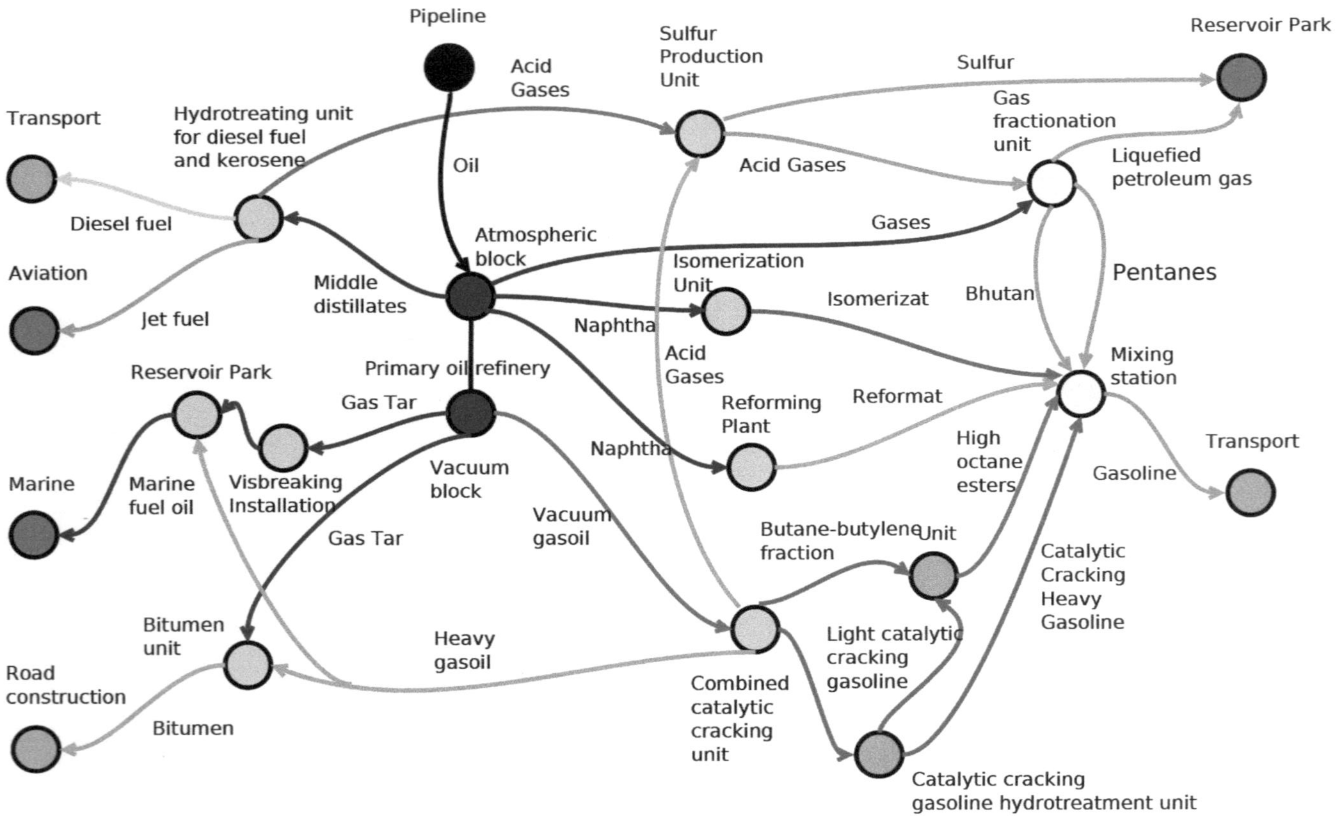

Fig. 1.16 Oil refinery model as a graph.

1.6 Management, control problems, and uncertainty

When designing the safety management and control systems of large-scale infrastructure, we need to solve some important issues as follows:

- Define the balanced set of global goals for effective management and control.
- Estimate possible levels of risk that we can achieve when manage and control.
- Calculate the overall cost of resources that are reasonably needed for achieving the delivered goals with a given performance.

Since the requirements of characteristics of distributed control systems for modern large-scale infrastructure are constantly strong, the complexity of management and control systems and the complexity of their design process and maintenance become the key factors determining the successes of design solutions and the efficiency of risk control.

The main idea behind controlling complex systems is to divide control tasks to some hierarchical levels (Mesarouic, Macko, & Takahara, 1970).

One way to achieve this is through the development of hierarchical intelligent control systems on the basis of the Increasing Precision with Decreasing Intelligence (IPDI) principle (Saridis, 2001).

The main ideas of the IPDI principle are analyzing control objectives and tasks, and joining them depending on the requirements to a precision of control and complexity of implementation of control tasks solution. So, we can say, the higher the required control precision for some hierarchical level is, the lower the requirements are to the control intelligence; and on the contrary, the higher the required level of intelligence is, the lower the precision of control is that is required. The universal measure of the complexity being used here is the entropy of management and control processes.

The application of such an approach makes it possible to estimate and decrease complexity of management decisions on safety provision for large-scale infrastructures. The general architecture of hierarchical management and control systems includes three vertical interacting levels: the organizational, the coordination, and the executive control level. The dependence of the complexity of the design of the control system on the complexity of the control object is based on the law of the required diversity. The main idea of this law is that a more complex system needs more complex control systems (Ashby, 1956).

The indicated tasks of creating the management and control systems for risk control provision are related to the class of multicriteria optimization problems. For solutions of the tasks mentioned, we need to use algorithms based on soft computing methods: fuzzy logic, neural networks, genetic algorithms, deep learning, image recognition, etc. All of these are based on the learning algorithms that use the set of examples to learn. We therefore need to select these examples carefully on the basis of big data technologies. When we design such sophisticated management and control systems, we need to apply some basic principles as follows:

- principle of functional integration which supposes the procedure of unification of different subsystems;
- principle of hierarchical organization, meaning the design of the control system in a class of multilevel hierarchical control systems with a division into several control levels differing by a choice of control goals and methods;
- principle of integrating various models and methods; and
- principle of open system construction, being the basis of intellectualization and standardization of information processing technologies at the various stages of the safety system life cycle.

Fig. 1.17 presents a block diagram of the multilevel intelligent control system of safety provision consisting of the executive level, the coordination level, and the organizational level.

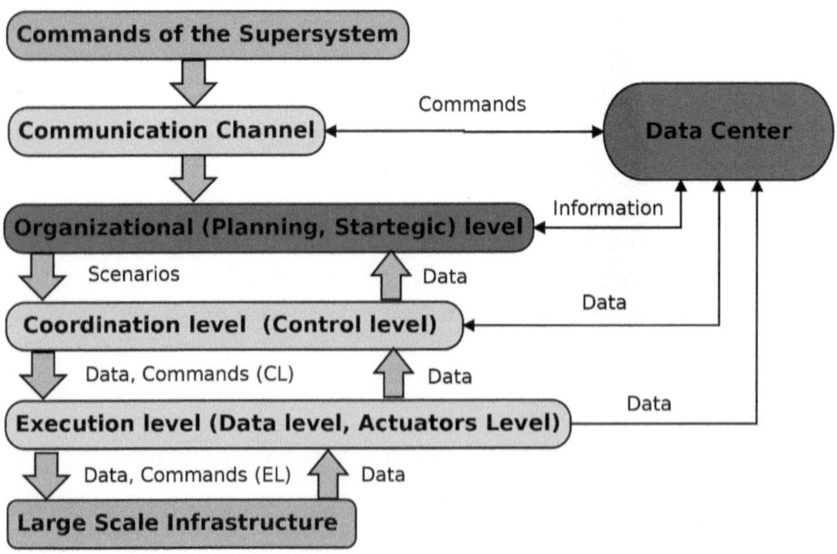

Fig. 1.17 Multilevel intelligent control system of safety provision.

In the following chapters, we will discuss the features of building a hierarchical management system and applying big data.

1.7 Personnel and process safety competency

Large-scale infrastructures may be considered as complex sociotechnical systems. System safety plays a key role for process industries belonging to this class of systems. Personnel and process safety competency is a basic issue while providing process safety for large-scale infrastructures.

A sociotechnical system (STS) is a set of interconnected material objects (technical means and personnel, ensuring their functioning and their use) intended for the direct execution of a system function. In other words, from a systemic point of view, an STS is a hierarchical human-machine complex that functions purposefully to realize its properties in accordance with its intended purpose. In our case, the purpose is providing process safety for the large-scale infrastructure.

The simplest (elementary) STS is a workplace (control room) with staff in it (dispatcher). In the general case, a real STS is a hierarchical system formed by a set of simplest (elementary) STS.

The STS organization is understood as a way of interconnection and interaction between its elements, ensuring their integration into this STS. The STS organization is divided into constant (invariant) and variable parts. The first part determines the structure of the STS and the second the program of its functioning. In the general case, the STS consists of two subsystems: the controlling and the managed, the elements of which implement the functions of converting resources into a result.

The process safety personnel deal with highly dangerous facilities and chemical components. Technicians work with these plants and components directly. They can create a dangerous situation or accident through incorrect action caused by lack of safety knowledge and safety competency. The engineers and managers are responsible for safety decision-making and safety strategy, meaning that they may also cause an emergency by making incorrect decisions.

Therefore, safety data are a critical resource for process safety management. The data have very complex structure and large volume, so the personnel are required to have a degree of competency to fulfill the tasks of process safety providing and managing.

Competency can be determined as an important skill necessary to do a job in general. In accordance with a Process Safety Management (PSM) program, this is an element related to efforts to maintain, improve, and broaden safety knowledge and expertise. The element provides the implementation of the body of process knowledge to manage risk and improve a facility's characteristics. The main outcome of safety competency is realizing and analyzing knowledge that helps, in the frame of a sociotechnical system, to provide appropriate decisions and ensure that personnel dealing with an emergency have enough skills and knowledge to act properly.

Process safety competency implies personnel continuing learning and training, including implementing big data technologies. The main source of the information for learning comes from process safety past incidents. Usually, companies implement information about incidents from one facility or from all the enterprise. Collecting and analyzing the incident data from different companies and countries on the basis of big data technologies would make the learning process more effective. In addition, a new aspect of refining training programs is implementing interactive software tools as well as virtual reality technologies.

Knowledge of process safety includes activities related to the compilation, cataloging, and provision of a specific data set, which is usually recorded in a paper or electronic format. Thus, employees of modern process industries should have the necessary knowledge and skills in the field of information technology, and a high information culture. Namely, personnel should understand the importance of safety data, be able to record it carefully, and prepare it for storage and analyzing, as well as being able to work with information systems. In the framework of big data technology, it is preferable to make electronic but not paper records.

Safety competency involves activities that help ensure the company collects and stores critical information on incidents. The required level of safety knowledge and information culture depends on the positions held by employees and their job functions or duties. An example of jobs, job functions, and necessary safety knowledge is presented in Fig. 1.18.

For instance, the awareness level requires minimal knowledge of process safety that is relevant to an employee's job functions. This means an employee should understand the hazards of the processes and plant that they deal with, and whom they can address to obtain additional information on a hazardous situation. This level of knowledge is common for operators, technicians, and contractors. They also need to have enough competencies for proper collecting and recording of information about plant functioning and functional parameters deviation.

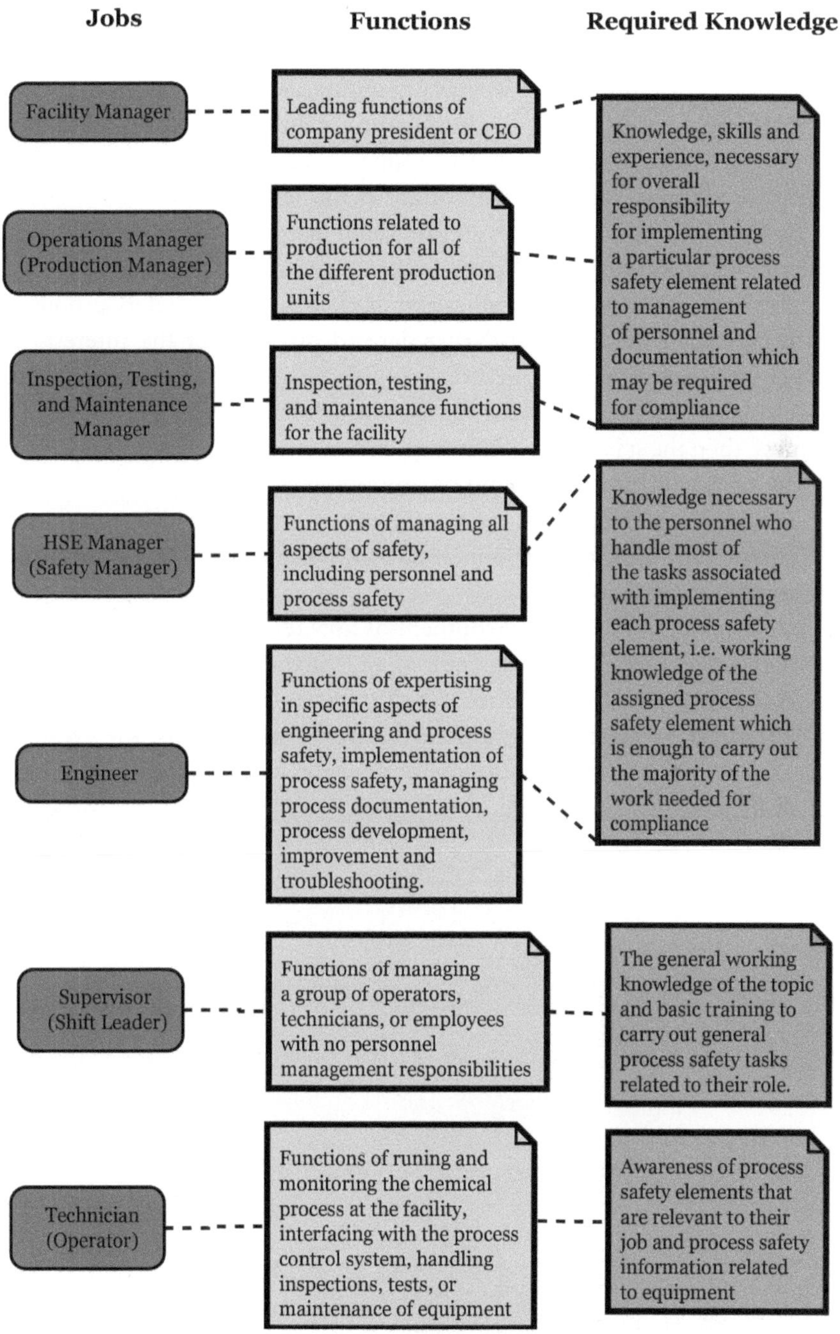

Fig. 1.18 The example of jobs, job functions, and necessary safety knowledge.

On the other hand, senior managers in the field of process safety of the entire enterprise require integrated knowledge in the field of safety of the entire production process, as well as organizational and technical measures for managing process safety. An information culture and understanding of the importance of big data analytics to support decision-making in the field of process safety are especially important at this level.

Stakeholders. Fig. 1.19 shows a stakeholder diagram. Next, we will try to understand the interests of all parties represented in this diagram.

The implementation and application of big data to solve safety problems in the industries we are considering depend on whether the interests of stakeholders coincide. The question is usually who owns the final product.

It can be assumed that the answer is obvious—this is industry. The next is business, then the scientific community, and, ultimately, society. If this is so, we can try to perform some kind of simple analysis:

- **Industry**: *benefits*: stable profit, development, competitiveness; *costs*: equipment, software, personnel, training, maintenance, rental of cloud services.
- **Business**: *benefits*: the development of new technologies, competitiveness, profit, new jobs; *costs*: equipment, software, personnel, training.
- **Academy and education**: *benefits*: new specialties, training and retraining of specialists, research, research funding; *costs*: the creation of laboratories, the development of new courses and educational programs.

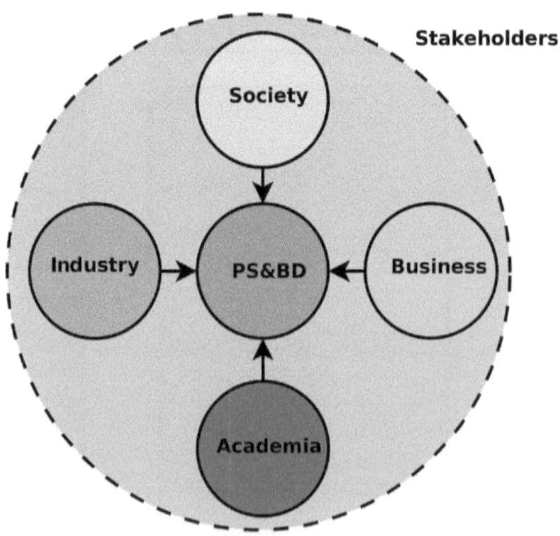

Fig. 1.19 Stakeholders set.

- **Society**: *benefits*: environmental safety and ensuring high standards of living for members of society, new jobs; *costs*: maintaining a business, developing standards and legal norms, monitoring compliance with standards.

1.8 Life cycle and safety problems

The concept of a life cycle in systems theory was introduced to analyze the development of complex infrastructures and their support. Thus, not only was the decomposition of a complex system into subsystems used, but also the development of the system over time was taken into account. The division into stages of the life of the system reflects the goals of each stage and its tasks that are solved to achieve the goal of the stage. A complex system is created by the labor of engineers for a long time. Therefore, the main task of optimizing the life cycle is the redistribution of time between stages in the direction of increasing the return on development costs due to the rapid implementation and effective operation of the designed system.

Of great importance for optimizing the life cycle is the family of standards governing the design, development, operation, maintenance, and disposal of a complex technical system. Safety standards include all stages of the life of a complex technical system. They reflect the experience and knowledge of engineers who are directly involved in all tasks arising at various stages of the life cycle, including the tasks of process safety.

Despite the fact that due attention is paid to safety issues at all stages of the life cycle of a complex system, from time to time, critical situations occur related to harm the human life and health, causing significant damage to the environment, as well as destruction of technological equipment. The reasons for these events are different.

1.8.1 Life cycle of large-scale infrastructures

The concept of a life cycle is used in the construction of information systems for complex technical objects, such as aircraft engines, aircraft, or satellites. This approach uses a method for analyzing complex information processes from the perspective of dividing the lifetime of complex systems, objects, and phenomena at certain stages. From the point of view of a systematic approach, we perform the decomposition of a complex phenomenon into stages. This is a relatively effective method for analyzing the characteristics of an object that change over time. Developments in this area are reflected in many standards and publications.

1.8.2 Energy life cycle

Fig. 1.20 shows the main stages of energy carrier life and possible connections between the stages. The main stages include extraction, processing, enrichment, storage, use, and disposal. The names of the stages reflect complex technological processes of various natures, the safety of which requires significant resources.

Throughout the entire life cycle, many different supporting processes are carried out, such as transportation. In this case, the name of the stage of the life cycle coincides with the process. Transportation safety is also a challenge.

At each stage of the life cycle of the energy carrier, a large amount of data is generated. It should be noted that these data can be used by the data sources themselves and provide the information needed at other stages of the energy carrier life cycle.

1.8.3 Data life cycle

We will use the concept of a life cycle in order to understand the dynamics of qualitative changes in data and their various states. Fig. 1.21 shows a diagram of changes in data states, where the generation stage includes many different ways and methods of translating the process characteristics into a measurement and comparison scale. The scales can be numerical and quality. During the generation process, data arrays are formed at some points in time. The

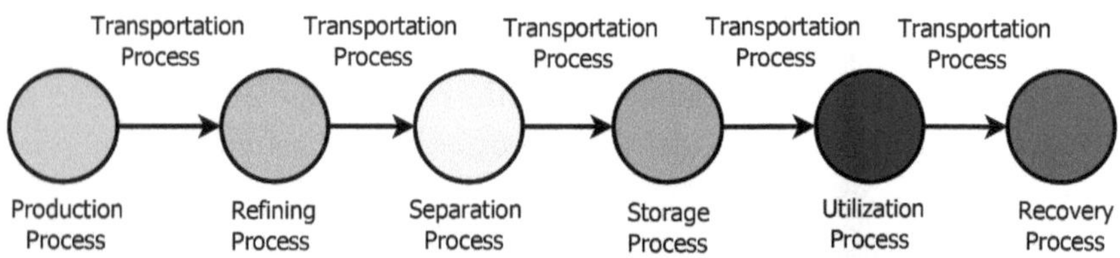

Fig. 1.20 Life cycle of energy carriers and their transportation.

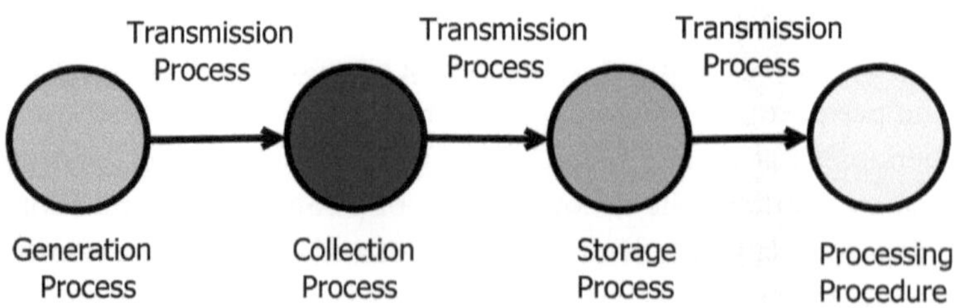

Fig. 1.21 Data life cycle.

processes of measuring and generating qualitative assessments will be discussed in Chapter 3. These processes characterize the stage of data collection.

Arrays of data are transmitted through various communication channels in the form of electrical signals, which are converted into digital form and recorded on various media. Data can be in a state of storage for a long time. Like everything related to storage, special conditions need to be created in order to save the data. If necessary, these data are processed and converted into other types of data representation. This process can be repeated many times.

The following chapters will discuss the features of the various stages of the data life cycle.

1.9 Standards and safety issues

A standard is a regulatory and technical document that establishes a set of norms, rules, requirements for an object of standardization, approved by a special body. In the broad sense of the word, a standard is a sample or model, taken as the starting point for comparison with other similar objects.

For the process industry, which is a highly hazardous area, it is especially important to follow the standards and regulations in the field of process safety. The regulations provide certain steps to follow for all the aspects of process safety, from the equipment technical requirements and operating procedures to the safety corporate culture and personnel safety competency. Regulations compliance helps to increase process safety and decrease process, operational, and environmental risks. At present, only a few countries have government process safety regulations, including the USA, EU, UK, Russia, Australia, Japan, and some other countries. These countries are the leaders in process safety standards development and propagation. The main institutions providing the standard in the process safety area are OSHA, CCPS, EU-OSHA, CIA, NOPSEMA, SWA, ISO, and IEC.

The **Occupational Safety and Health Administration** (OSHA) belongs to the U.S. Department of Labor. OSHA's mission is to "assure safe and healthy working conditions for working men and women by setting and enforcing standards and by providing training, outreach, education and assistance" (*OSHA*, 2020). The agency is also charged with enforcing a variety of whistleblower statutes and regulations. The Occupational Safety and Health Act granted OSHA the authority to issue workplace health and safety regulations. These regulations include limits on hazardous chemical exposure, employee access to hazard information, requirements for the use of personal protective equipment, and requirements to prevent falls and hazards from operating dangerous equipment. OSHA is responsible for enforcing its standards on regulated entities. Compliance Safety and Health Officers carry out inspections and assess fines for regulatory violations. Inspections are planned for worksites in particularly hazardous industries. Inspections can also be triggered by a workplace fatality, multiple hospitalizations, worker complaints, or referrals. Tracking and investigating workplace injuries and illnesses play an important role in preventing future injuries and illnesses. Under OSHA's recordkeeping regulation, certain covered employers in high hazard industries are required to prepare and maintain records of serious occupational injuries and illnesses. This information is important for employers, workers, and OSHA in evaluating the safety of a workplace, understanding industry hazards, and implementing worker protections to reduce and eliminate hazards. OSHA has developed several training, compliance assistance, and health and safety recognition programs throughout its history.

The **Center for Chemical Process Safety** (CCPS), part of the American Institute of Chemical Engineers (AIChE), is a nonprofit, corporate membership organization that deals with process safety in the chemical, pharmaceutical, and petroleum industries. It is a technological association of manufacturing companies, government agencies, educational institutions, insurance companies, and experts aiming to improve industrial process safety. CCPS publishes numerous guidelines in the field of process safety (*CCPS*, 2020).

The **European Agency for Safety and Health at Work** (EU-OSHA) (*EU-OSHA*, 2020) is a regionalized agency of the European Union with the job of collecting, analyzing and distributing appropriate information intended to benefit people involved in safety and health at work. The EU-OSHA collects, analyzes, and disseminates safety and health information throughout the EU and provides the evidence base that can be used to advance safety and health policies. EU-OSHA publishes monthly newsletters, which address occupational health and safety problems, and also provides detailed reports on occupational safety and health data.

The **Chemical Industries Association** (CIA) (CIA, 2019) is the leading national trade association representing and advising chemical and pharmaceutical companies located across the United Kingdom. The CIA represents member companies at both national and international level. The CIA carries out advocacy on behalf of its members' interests. Its remit is to "articulate members' collective hopes and concerns; improve appreciation of the situation amongst UK and European governments, the media and other key stakeholders; and help address the global competitive issues encountered by our members."

The **National Offshore Petroleum Safety and Environmental Management Authority** (NOPSEMA) is Australia's independent expert regulator for health and safety, environmental management, structural, and well integrity for offshore petroleum facilities and activities in Commonwealth waters (*NOPSEMA*, 2020).

Safe Work Australia (SWA) is an Australian government statutory body established in 2008 to develop national policy relating to work health and safety workers' compensation (Safe Work Australia (SWA), n.d.).

The **International Organization for Standardization** (ISO) is an international standard-setting body composed of representatives from various national standards organizations (ISO, 2020). ISO defines its tasks as follows: the development of standardization and related activities in the world to ensure the exchange between users and services, as well as the development of cooperation in the intellectual, scientific, technical, and economic fields. ISO is an organization that develops international standards, but it does not carry out conformity assessment and/or certification for compliance with standards. In this regard, ISO never issues certificates and no company can be certified by ISO. Certification is carried out by independent certification bodies; the largest and most famous are TÜV (Austria and Germany), BSI (Great Britain), BVC (France), DNVGL (Norway), SGS (Switzerland), ABS (USA), and LRQA (Great Britain). ISO developed ISO 31000:2018, Risk management—Guidelines providing the main principles, framework, and process for managing risk. These guidelines can be used by any company, regardless of its size, type of activity, or area, in terms of economic indicators and professional reputation, as well as environmental, safety, and social results.

The **International Electrotechnical Commission** (IEC) is an international nonprofit organization for standardization in the field of electrical, electronic, and related technologies (*IEC*, 2020). Some of the IEC standards are being developed in collaboration with the International Organization for Standardization (ISO), such as IEC 31010:2019, Risk management—Risk assessment techniques. This is one of the key standards for the risk assessment in different areas including safety. The standards ISO 31000:2018 and IEC 31010:2019 will be analyzed in Chapter 2.

As we can see, the main function of the listed above institutions are monitoring and enforcing standards compliance as well as investigation of accidents. Another important function of these institutions is promoting standards and regulations, and advising to improve their implementation efficiency.

Most of the standards and regulations in process safety area are dedicated to the Process Safety Management (PSM) and Risk Assessment methods. The most significant standards, regulations, and methods will be considered in detail in the following chapters. To illustrate the process safety documentary life cycle and big data's possible role, let us consider the main components of PSM usual for PSM programs developed by different international organizations:

- management system;
- employee participation;
- process safety information;
- process hazard analysis;
- operating procedures;
- personnel training;
- contractors;
- prestartup safety review;
- mechanical integrity testing and training;
- hot work permit;
- management of change;
- incident investigation; and
- emergency planning and response compliance audits.

These components are comprehensive and interconnected, and require the following data and documentation in a certain form.

Process hazard analysis (PHA) (or process hazard assessment) is a collection of organized and systematic assessments of potential hazards associated with a manufacturing process. PHA provides information designed to help managers and employees make decisions to improve safety and reduce the effects of unwanted or unplanned releases of hazardous chemicals. PHA seeks to analyze the potential causes and consequences of fires, explosions, toxic or flammable chemicals, and major spills of hazardous chemicals, and focuses on equipment, appliances, utilities, human activities, and external factors that may affect the process.

A *prestartup safety review* (PSSR) is a safety check carried out before the start-up (commissioning) of a new or modified processing/production unit or installation, to ensure that the plants comply with the original design or operational plan, in order to catch and re-evaluate any potential hazard due to changes during detailed engineering and construction phase of the project.

The *process safety information* is available and complete for all the systems including *material safety data sheets* (MSDS) and other documentation (Process Safety Management of Highly Hazardous Chemicals – Compliance Guidelines and Enforcement Procedures).

Standard operating procedures (SOPs) are written for each system, and include appropriate security procedures and other relevant information.

The *mechanical integrity program* (MIP) is designed to ensure that key equipment is properly maintained, including pressure vessels and storage tanks, piping and ventilation systems, control systems, pumps, and emergency shutdown systems.

Piping and instrumental diagrams (P&IDs) perform every process component, and include make and model.

Documentation on *management of change* (MOC) indicates that every change made to a process has been evaluated for its impact on safety (Aziz, Shariff, & Rusli, 2014).

Training schedules and documentation (TRAIN) show that every employee has passed through the proper training.

All these groups of data and documents have different structures and natures (diagrams and drawings, tests results, text files, scanned handwritten documents, etc.) (Guidelines for process safety documentation, Center for Chemical Process Safety, 1995). Herewith, the tasks of big data technologies are collection and processing of heterogeneous data, classification, analysis, modeling, and forecasting as well as delivering necessary information by request to numerous users in the real-time mode.

1.10 Process safety and big data

As was mentioned before, process safety is also associated with the processing of statistical data that are used in calculating risks and developing a set of detailed scenarios of response to a critical situation.

When developing mathematical models of a critical situation, the data obtained during the measurement of technological process parameters in time are widely used, and the modeling results can also serve as an additional source of important information.

The development of modern technologies for the data collection, network and radio transmission, storage in data centers, and processing of large amounts of data (big data) allows us to solve some urgent problems associated with identifying the characteristic ways of critical situations arising in various industries: oil and gas, petrochemicals, and others.

These data can be used to clarify the safety standards in the field of hazard technological processes, for more effective training of personnel and the development of various models to avoid critical situations.

It is important to understand that for effective implementation of big data technologies while ensuring process safety issues, additional costs are required, along with identifying the set of stakeholders, recruiting highly qualified personnel, and developing new standards in this area.

1.10.1 Big data phenomenon

Data collection and processing of analytical reports are common practices in any company. Data and their processing are taken quite seriously (Mohammadpoor & Torabi, 2018).

We will briefly consider the evolution of the usual data processing using spreadsheets. Processing data using spreadsheets allows the desired results to be obtained if the processed data do not exceed a certain amount. On a personal computer, it is impossible to process an array of data exceeding the critical size associated with the amount of RAM in the computer. In this case, we are faced with a restriction on the amount of processed data. A solution was found, and we began to use specialized software and specialized servers for storing and processing significant amounts of data: so-called databases. These databases processed large amounts of data, which allowed us to move on to the procedures for selecting the necessary information upon request.

We are now living in an era of data. Around us there are many sources of information that can be represented as data sets. We are trying to store more and more information, because our storage systems in the private cloud are now more cost-effective in terms of storing data sets. Database systems became more complex, and there was a need for their maintenance and improvement, which required the participation of programmers and system administrators in the data life cycle. The transition from reports to analytics required the involvement of analysts who could explain the facts from the reports and build a long-term forecast.

It should be noted that the amount of data processed requires the involvement of several categories of specialists with different competencies in their processing. In turn, to manage this group, it was necessary to develop management technologies in the spirit of Agile; as a result, changes are taking place in the composition of the organization's personnel and in the methods of developing management decisions based on data. Such organizations are called organizations in which data-based management is applied.

When we are trying to define big data as a phenomenon, the following features of it are noted:

- velocity;
- volume; and
- variety.

If we are talking about the *velocity* of data generation, we take into account the need to use an increasing number of devices for storing the data. Here the tasks of scaling in real time arise, i.e., increasing data storage system size in nonstop modes.

When talking about the *volume* of stored data, there is a need to take into account the problems associated with storage for data arrays, their aging, and loss of relevance. We need to rewrite the data periodically, as they are in digital form on different types of physical data storage devices.

When *variety* is noted, it means that we have to process scanned copies of documents, text, data from various sensors, digital photos, video films, and audio files. All these forms of electronic documents have their own storage formats and various algorithms for their processing. We therefore have the some problems to process all these data as an integrated data structure.

Big data issues are studied in the scientific fields of data science, computer science, and artificial intelligence. We are trying to find the answer to the question of how to apply big data technologies effectively to ensure the effective solutions of process safety tasks provision.

The peculiarity of this subject area is its multidisciplinarity. The core of this subject includes knowledge from various fields of human activity: process safety, design and operation of technological equipment, process control, applied and theoretical informatics, and information technology. As we know, computer science is associated with algorithms for collecting, transmitting, storing, and identifying dependencies based on data. Information technologies include the development and use of various programming languages and software engineering. Applied computer science focuses on the features of the industry for which algorithms and information systems are being developed.

Large arrays of digitized data are stored in data centers. The processes of storing data arrays require significant material costs. At the same time, a large number of servers, special software, and high-speed communication channels are used. For the operation of servers and cooling systems, a large amount of electricity is used. To solve reliability problems, data are duplicated, which requires a corresponding increase in costs. Big data technologies are a set of software tools and methods for processing structured and unstructured data of large volume, presented in different formats: text, graphic, etc. The Big Data Toolkit includes mass-parallel data processing tools based on NoSQL technologies, MapReduce algorithms, software frameworks, and Hadoop project libraries. As a sidenote, it was believed that traditional relational database management systems (DBMSs) based on SQL query language were not suitable for processing large amounts of information. However, modified versions of relational databases can now be used successfully to work with big data (Gorman, 1991).

Today, big data technologies are becoming the basis for creating surplus value, and in some cases can improve technological processes and process safety (Mohammadpoor & Torabi, 2018; Ouyang, Wu, & Huang, 2018). In industrially oriented countries various projects have been started for research and application the big data technologies, for instance, in the United States, the 'Big Data Research and Development Initiative' in 2012, and in 2016 "The Federal Big Data Research and Development Strategic Plan" was released. In China, the Ministry of Industry and Information Technology has developed a big data infrastructure development plan using standardized systems. In Japan, big data technologies have been a key component of the national technology strategy since 2012. The United Nations (UN) launched the Global Pulse Initiative in 2009 to collect big data for humanitarian development.

1.10.2 Big data and safety problems

The procedures for collecting information about incidents that have occurred during the execution of technological processes are regulated by various standards and internal regulations at the enterprise (*OSHA*, 2020; CIA, 2019).

Information about incidents that have occurred at the enterprise is recorded in special documents in paper or electronic form by safety specialists or responsible persons. These documents have a specified standard view with all the points for entering data when describing incidents.

The procedure for analyzing the causes of the incident and generating reports with conclusions are reflected in various standards that determine the main stages of the investigation of the incident (*OSHA*, 2020).

Companies are required to post incident information in official releases and open access databases.

Fig. 1.22 shows the process of generating a flow of data on the state of technological processes, failure analysis, and incident analysis, where:

- *DB_P* is the database of the results of personnel monitoring the progress of technological processes;
- *DB_S* is a database of measurement results of the characteristics of technological processes performed by sensors;
- *DB_f* is a database with the results of failure analysis and equipment failure and violations during technological processes; and.
- *DB_h* is a database with information about critical situations and their consequences.

Part of the data is reflected in various forms, such as reports presented in paper documents. This information is then entered into the relevant information systems and stored in databases.

For example, process control reports contain information about parameter deviations and their causes. These reports are generated in two ways: by a technologist and using an automated control and monitoring system (SCADA).

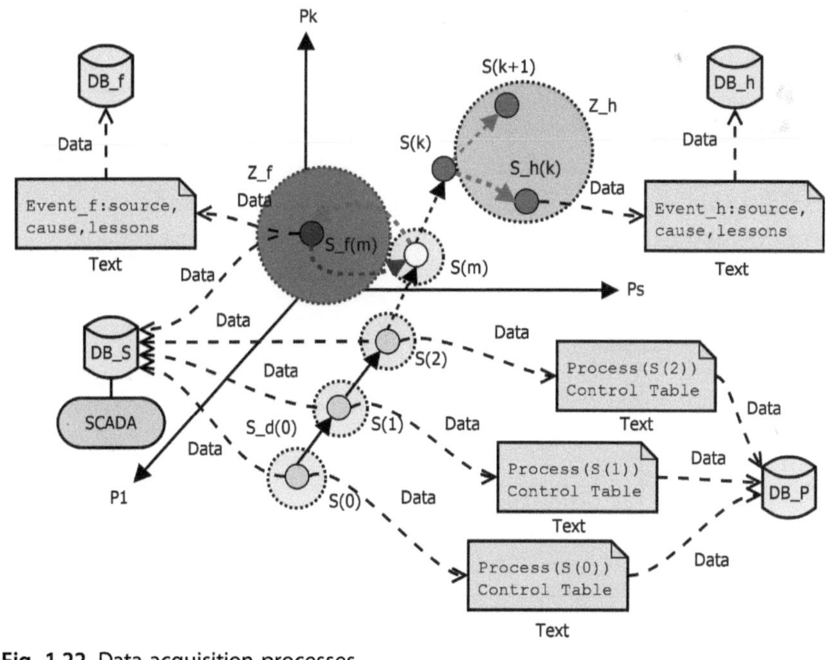

Fig. 1.22 Data acquisition processes.

When implementing various technological processes at various facilities, a large data array is formed that reflects the current state of the large-scale infrastructure. The complexity of technological processes and the requirement of their constant monitoring lead to the fact that an extremely large data array is formed, which must be stored for a given time. Often the data is stored in different unrelated information systems, which in some cases does not allow us to analyze the occurrence of a critical situation taking into account all available data.

Arrays of data, of course, are relevant if it is necessary to take into account certain features of the process, the state of the environment, and possible violations of safety standards by personnel. These arrays are invaluable information assets of the enterprise. When developing forecasts and digital counterparts, these data stored in databases certainly help to increase the efficiency and quality of required solutions (see Fig. 1.23).

Active work is underway to introduce big data technologies to solve the problems of ensuring the safety of technological processes. However, given the variety of large infrastructures and their complexity, solving these problems requires a great deal of effort (Valeev & Kondratyeva, 2016a, 2016b).

Fig. 1.24 shows the process of generating a flow of data on the state of technological processes, failure analysis, and incident analysis, where:

- *DB_P* is the database of the results of personnel monitoring the progress of technological processes;
- *DB_S* is a database of measurement results of the characteristics of technological processes performed by sensors;
- *DB_f* is a database with the results of failure analysis and equipment failure and violations during technological processes; and
- *DB_h* is a database with information about critical situations and their consequences.

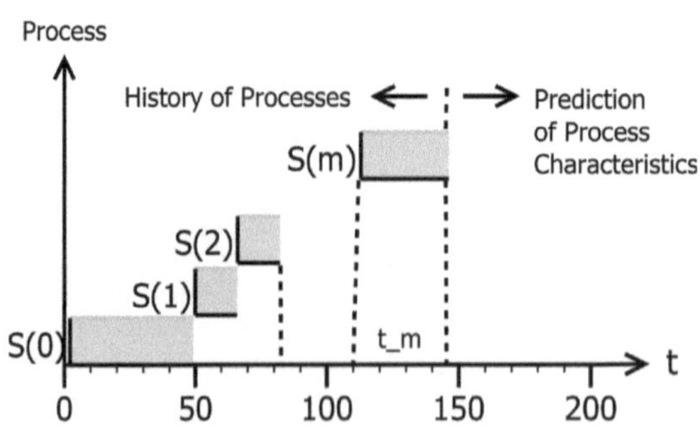

Fig. 1.23 Time diagram of technological processes.

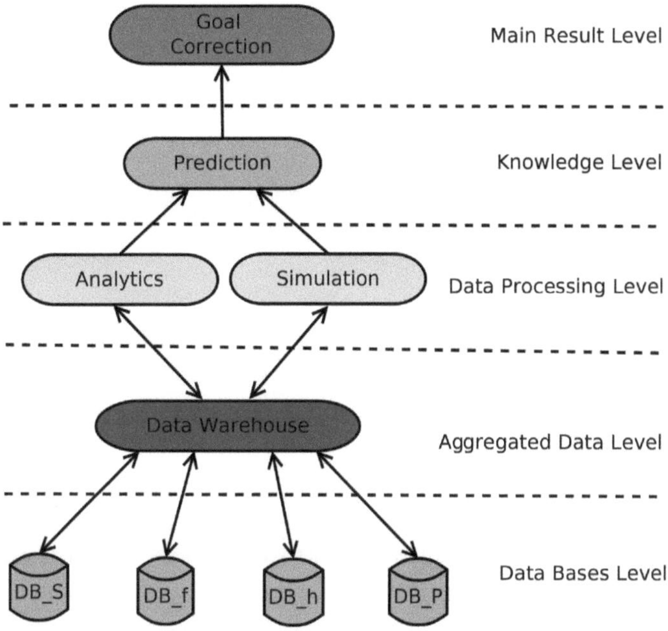

Fig. 1.24 Data flow from databases level to main result level.

The use of big data technologies allows implementing the hierarchical data flow control system. Data collected in various databases are aggregated in a data warehouse. At the level of processing aggregated data, problems of analysis and modeling of the state of the enterprise or company are solved. The results obtained at this level are used to develop a forecast of the state of the main indicators of enterprise performance, including indicators related to the safety of technological processes. At the highest level, based on the forecast obtained, the problem of adjusting the goals of the company can be solved.

1.10.3 Roadmap of big data implementation for process safety

Taking full advantage of big data technologies to ensure process safety requires a high data culture in the company. The most important thing in the formation of this culture is a careful attitude to the information assets available to the organization. Data collection requires some effort and investment. However, it is not always clear which data are useful, how long the data should be stored, and most importantly, how the data should be used later. This long journey from data collection to its use needs to be covered, and time will put everything in its place.

Furthermore, we need to consider how we see the main stages of the implementation of big data technologies to ensure technological safety at the enterprise.

1.10.3.1 Step 1. Data audit of infrastructure

Main goal: Perform an audit of the available data responsible for improving the performance of safety functions in accordance with standards and regulations.

Main procedures of realization:

1. Make a list of safety issues for objects and infrastructure and decisions made for their implementation.
2. Make a list of the necessary data and their sources for the effective performance of safety issues.

Notes:

1. Typically, data are collected to generate standard reports, i.e., the goal in this case is to generate a report. After that, it is not clear what to do with the data arrays. They can be left in storage, or they can be deleted. When storing, it is not always clear how long the data need to be stored.
2. The audit process raises many questions related to the storage time, value of information, and its aging. Currently, information storage has become part of the data processing industry, and these issues are being addressed within the cloud storage by companies from the information industry. When storing data in the storage systems of contractors, in turn, the task arises of maintaining the confidentiality of information resources.

1.10.3.2 Step 2. Data classification

Main goal: Classify availability of data and their sources.

Main procedures of realization:

1. Determine the availability of data with known sources.
2. Define a list of necessary data that are lost or possibly lost.
3. Define a list of missing required data with known sources.
4. Define a list of data that you need, but their sources are not available, or unknown.

Notes:

1. Data mining requires the organization's resources. If these are sensors that determine the concentration of harmful substances, then when choosing the type, quantity, and places of their installation, it is necessary to remember the data life cycle, as mentioned earlier. Maintenance of the data life cycle requires compliance with metrological standards,

certification of sensors and software, certification of communication channels, and information protection systems.

2. Note that when using big data, information resources of other companies can be used. In this case, the question arises about the quality of data provided by a third-party company.

1.10.3.3 Step 3. Classification of data by priority

Main goal: Prioritize and highlight for the most important data resources.

Main procedures of realization:

1. Assess the complexity of obtaining data.
2. Make a list of data priority and costs of their collection.
3. Take into account the dynamics of priorities in the process of performing the safety functions.

Notes:

1. When performing this step, you should be aware of the contradictions that may arise when choosing a priority scale. The choice of priorities is a difficult task to solve within the framework of the theory of decision support systems.
2. The cost of obtaining data depends largely on the type of infrastructure and its location.
3. Costs may vary due to lower costs for sensors and microcomputers.

1.10.3.4 Step 4. Analysis of data acquisition methods

Main goal: Improve data efficiency.

Main procedures of realization:

1. The labor function of personnel should indicate how to collect and work with data.
2. Time-consuming operations must be performed by dedicated personnel.
3. It is necessary to ensure effective interaction with this staff.

Notes:

1. Despite the active introduction of automated data collection systems, manual operations remain for monitoring the state of structures, for example, external inspection of the installation, pipeline, welds, etc.
2. As part of the implementation of a data culture, it is important to understand that data collection requires the participation of qualified specialists at various levels.

1.10.3.5 Step 5. Determine the method and place of data storage

Main goal: To provide reliable data storage and access on demand.

Main procedures of realization:
1. Use cloud data storage.
2. Provide data protection measures based on the distribution of access rights.
3. Control changes and versioning.

Notes:
1. As follows from the presented basic steps of working with data or the formation of a data culture, this is a rather time-consuming process.
2. Currently, modern companies are introducing the position of data officer and data manager. Note that these are the first basic steps toward applying big data.

In the following sections of the book, we will consider:
- data sources and methods for their collection;
- multilevel systems of management and control of process safety in infrastructures;
- modern standards related to the theme of the book;
- information technology, which is the basis for the implementation of big data;
- algorithms and methods for developing technological safety models; and
- algorithms and methods of big data analytics.

1.11 Summary

We reviewed the concepts used in systems analysis to understand process safety management to minimize risks using big data technologies. It was mentioned that industrial infrastructures are located on land or in water, some can change their location, and all of them have a number of typical features that characterize them as a special class of complex systems. They have common features: they occupy a huge area and volume, and their organizational structure includes many interconnected complex technical objects.

The functioning of a large infrastructure is supported by a large group of people of various qualifications and professional knowledge. Large-scale infrastructures can be dangerous to people and the environment. To find ways to reduce risks in emergencies, we discussed the nature of the complexity of infrastructures. This is necessary in order not to give the impression of the impossibility of monitoring the state of the infrastructure, but on the other hand, to aid understanding of the price that must be paid for safety.

When we design the safety management and control systems of large-scale infrastructure, we need to solve some important issues as follows:

- Define the balanced set of global goals for effective management and control.
- Estimate possible levels of risk that we can achieve when manage and control.
- Calculate the overall cost of resources that are reasonably needed for achieving the delivered goals with a given performance.

Large-scale infrastructures may be considered as complex sociotechnical systems. Personnel and process safety competency are the basic issues while providing process safety for large-scale infrastructures.

The concept of a life cycle in systems theory was introduced to analyze the development of complex infrastructures and their support. Thus, not only was the decomposition of a complex system into subsystems used, but also the development of the system over time was taken into account. The division into stages of the life of the system reflects the goals of each stage and its tasks that are solved to achieve the goal of the stage. A complex system is created by the labor of engineers for a long time. Therefore, the main task of optimizing the life cycle is the redistribution of time between stages in the direction of increasing the return on development costs due to the rapid implementation and effective operation of the designed system.

The safety standard is a sample model, taken as the starting point for analyzing safety issues. Safety is based on the mandatory implementation of various rules laid down in safety standards. Modern safety standards reflect our current views on the use of modern information technologies that allow for the collection, transmission, storage, and processing of significant amounts of data.

Large objects utilize various sources of information that allow them to be represented in digital form: various sensors of the state of technological equipment; the state of the external environment; and video sensors recording the main stages of human-machine operations. Data transmission requires the use of high-speed, noise-protected data channels. During the life cycle, a large amount of data is generated in the system. The data collected in data warehouses require the problems of their reliable storage and processing to be solved.

We are now living in an era of data. There are many information sources around us, which can be represented in the form of data sets. We are trying to store more and more information, since our private cloud store systems are being more cost-effective all the time.

Currently, active work is underway to introduce big data technologies to solve the problems of ensuring the safety of technological processes. However, taking into account the variety of infrastructures and their complexity, the solution of these problems requires quite a lot of effort.

Taking full advantage of big data technologies to ensure process safety requires a high data culture in the company. The most important thing in the formation of this culture is a careful attitude to the information assets available to the organization. Data collection requires some effort and investment. However, it is not always clear which data are useful, how long the data should be stored, and most importantly, how the data should be used later. This long journey from data collection to its use needs to be covered, and time will put everything in its place.

1.12 Definitions

A *system* is a group of interacting or interrelated entities that form a unified whole.

The *complexity* of the system is the function $C = f(n, V, p(n), q(V))$, where n is the number of elements, V is the connection between the elements, p is the procedure for determining the state of an element, and q is the procedure for determining the characteristics. Complexity denotes the behavior of a system or model, the many components of which can interact with one another in different ways, only following local rules but not instructions from higher levels.

A *sociotechnical system* is a set of interconnected material objects (technical means and personnel, ensuring their functioning and their use) intended for the direct execution of a system function.

A *standard* is a normative document or establishing rules, characterized by the features of development, approval, and methods of use, as well as focus on special purposes. In the broad sense of the word, a standard is a sample or model, taken as the starting point for comparison with other similar objects.

A *process industry* is an industry, such as the chemical or petrochemical industry, that is concerned with the processing of bulk resources into other products.

Process safety is the state of protection of the vital interests of the individual and society from accidents at hazardous production facilities and the consequences of these accidents. The main focus is on the prevention of fires, explosions and accidental releases of chemicals at chemical production facilities or other facilities associated with hazardous materials such as refineries and oil and gas production facilities (on land and at sea).

Large-scale industrial infrastructures are located on land or in water, some of them can shift their location, and all of them have a number of typical features that characterize them as a special class of complex systems: they occupy a large area and volume; their organizational structure includes many interconnected complex technical objects; and the functioning of the large-scale infrastructure is supported by a large group of people organized hierarchically.

A *data center* is a specialized building for hosting server and network equipment and connecting subscribers to internet channels. The data center performs the functions of processing, storage, and distribution of information, as a rule, in the interests of corporate clients. The consolidation of computing resources and data storage facilities in the data center enables a reduction in the total cost of ownership of IT infrastructure due to the possibility of efficient use of technical means, for example, redistribution of loads, as well as by reducing administration costs.

A *graph* is a set of points, called nodes or vertices, that are interconnected by a set of lines called edges.

A *hierarchical control system* is a form of management system in which a set of devices and management software are located in a hierarchical tree.

Safety data are the data related to all the aspects of the process safety concerning the use of various substances and products, process equipment, and facilities as well as the information on organizational aspects of providing process safety.

Process safety competency is related to efforts to maintain, improve, and broaden safety knowledge and expertise. It provides the implementation of the body of process knowledge to manage risk and improve facilities characteristics. The main outcome of safety competency is realizing and analyzing knowledge that helps, in the frame of a sociotechnical system, to provide appropriate decisions and make sure that personnel dealing with an emergency have enough skills and knowledge to act properly.

A *stakeholder*, also described as an interested party, involved party, work participant, or role in the project, is a person or organization who has rights, share, requirements, or interests regarding the system or its properties that satisfy their needs and expectations.

The *life cycle of a system* is the process stage, covering various states of the system, starting from the moment it becomes necessary in such a system and ending with its complete decommissioning.

The *data life cycle* is the sequence of stages that a particular unit of data goes through, from its initial generation or capture to its eventual archival and/or deletion at the end of its useful life.

Big data is the designation of structured and unstructured data of huge volumes and significant variety, effectively processed by horizontally scalable software tools, alternative to traditional database management systems and solutions of the business intelligence class. In a broad sense, "big data" is spoken of as a socioeconomic phenomenon associated with the advent of technological capabilities to analyze huge amounts of data. In some problem areas, this is the entire world volume of data, and the transformational consequences arising from this.

References

Ashby, W. R. (1956). *An introduction to cybernetics*. London: Chapman & Hall Ltd.

Aziz, H. A., Shariff, A. M., & Rusli, R. (2014). Managing process safety information based on process safety management requirements. *Process Safety Progress, 33*(1), 41–48. https://doi.org/10.1002/prs.11610.

Benintendi, R. (Ed.). (2018). *Process safety calculations* Elsevier.

Brennan, D. (2012). *Sustainable process engineering: Concepts, strategies, evaluation and implementation*. CRC Press.

CCPS. (2020). *Center for Chemical Process Safety (CCPS) website*. https://www.aiche.org/ccps.

Center for Chemical Process Safety. (April 15, 1995). *Guidelines for process safety documentation*. Center for Chemical Process Safety of the American Institute of Chemical Engineers.

CIA. (2019). *Chemical Industries Association (CIA) website*. https://www.cia.org.uk/ (Accessed 15 April 2020).

Curry, D. M., Beaver, W. W., & Dagli, C. H. (2018). A system-of-systems approach to improving intelligent predictions and decisions in a time-series environment. In *2018 13th System of systems engineering conference, SoSE 2018*. United States: Institute of Electrical and Electronics Engineers Inc. https://doi.org/10.1109/SYSOSE.2018.8428744.

EU-OSHA. (2020). *The European Agency for Safety and Health at Work (EU-OSHA) website*. https://osha.europa.eu/en. (Accessed 20 April 2020).

Gorman, M. M. (1991). *Database management systems*. Butterworth-Heinemann. https://doi.org/10.1016/B978-0-7506-0135-1.50002-6.

IEC 31010:2019. (2019). *Risk management—Risk assessment techniques*. ISO. Retrieved from https://www.iso.org/standard/72140.html.

IEC. (2020). *International Electrotechnical Commission (IEC) website*. https://www.iec.ch/ (Accessed 30 March 2020).

ISO. (2020). *International Organization for Standardization (ISO) website*. https://www.iso.org/home.html. (Accessed 2 April 2020).

Johnson, W. (1973). *The management oversight and risk tree*. Atomic Energy Commission.

Khan, F., & Abbassi, R. (2018). Methods in chemical process safety. *Offshore Process Safety*. Elsevier.

Leveson, N. G. (2016). *Engineering a safer world. Systems thinking applied to safety* (reprint ed.). The MIT Press.

Li, H. (2016). Communications for control in cyber physical systems. In *Theory, design and applications in smart grids* (pp. 1–294). United States: Elsevier Inc. Retrieved from http://www.sciencedirect.com/science/book/9780128019504.

Mannan, S. (2013). *Lees' process safety essentials: Hazard identification, assessment and control.* Butterworth-Heinemann.

Mesarouic, M. D., Macko, D., & Takahara, Y. (1970). *Theory of hierarchical multilevel systems.* Academic Press.

Mohammadpoor, M., & Torabi, F. (2018). Big data analytics in oil and gas industry: An emerging trend. *Petroleum.* https://doi.org/10.1016/j.petlm.2018.11.001.

Mokhatab, S., Mak, J. Y., Valappil, J. V., & Wood, D. A. (2014). *Handbook of liquefied natural gas.* Gulf Professional Publishing.

Mokhatab, S., Poe, W., & Mak, J. (2015). *Handbook of natural gas transmission and processing: Principles and practices* (3rd). Gulf Professional Publishing.

NOPSEMA. (2020). *National Offshore Petroleum Safety and Environmental Management Authority (NOPSEMA) website.* https://www.nopsema.gov.au/ (Accessed 10 April 2020).

Nriagu, J. O. (2011). *Encyclopedia of environmental health.* Elsevier Science.

Onwubolu, G. C. (2005). *Mechatronics.* Butterworth-Heinemann. https://doi.org/10.1016/B978-075066379-3/50002-1.

OSHA. (2020). *Occupational Safety and Health Administration (OSHA) website.* https://www.osha.gov/ (Accessed 30 March 2020).

Ouyang, Q., Wu, C., & Huang, L. (2018). Methodologies, principles and prospects of applying big data in safety science research. *Safety Science, 101,* 60–71. https://doi.org/10.1016/j.ssci.2017.08.012.

(1994). https://www.osha.gov/enforcement/directives/cpl-02-02-045-revised. (Accessed 18 March 2020).

Safe Work Australia (SWA) (n.d.). Retrieved January 2020, from https://www.safeworkaustralia.gov.au.

Sanders, E. R. (2015). *Chemical process safety: Learning from case histories* (4th). Butterworth-Heinemann.

Saridis, G. N. (2001). *Hierarchically intelligent machines.* World Scientific.

System (2020) . Retrieved from https://www.merriam-webster.com/dictionary/system.

Tekin, T., Pleros, N., Pitwon, R., & Hakansson, A. (Eds.). (2017). *Optical Interconnects for data centers* Woodhead Publishing.

Valeev, S., & Kondratyeva, N. (2016a). Distributed information and control system for emergencies in critical infrastructures. In *Presented at the 2016 IEEE 10th international conference on application of information and communication technologies (AICT).* Baku: IEEE. https://doi.org/10.1109/ICAICT.2016.7991653.

Valeev, S., & Kondratyeva, N. (2016b). Large scale system management based on Markov decision process and big data concept. In *Presented at the 2016 IEEE 10th international conference on application of information and communication technologies (AICT).* Baku: IEEE. https://doi.org/10.1109/ICAICT.2016.7991829.

WHO. (September 7, 2013). *Health risk assessment from the nuclear accident after the 2011 Great East Japan Earthquake and Tsunami.* World Health Organization.

Wu, C., & Buyya, R. (2015). *Cloud data centers and cost modeling: A complete guide to planning, designing and building a cloud data center.* Morgan Kaufmann.

CHAPTER 2

Risk and process safety standards

2.1 Risks and safety

One of the main areas of application of big data in the field of process safety is the analysis of process risks in order to reduce them. On the other hand, risk assessment is a necessary part of process safety management. Risk analysis is the procedure that helps in understanding the nature of risks and their features, including, when necessary, the level of risk. It also allows us to consider the specifics of the development of various types of accidents, as well as the determination of quantitative indicators of the associated social, material, and environmental damage.

2.1.1 Probability basics

The concept of probability plays an important role in assessing potential hazards and risks in the context of process safety management. In this section, we consider the main points of probability theory that are necessary for understanding the mathematical models and methods of risk theory, as discussed in subsequent sections and chapters. First we give definitions of event, experiment, and probability (Pugachev, 1984; Rumshiskii, 1965).

An *event* is the result of an experiment. An event is called random if it may or may not occur in a given experience. Let us designate the events with letters A, B, C, D, etc. By an *experiment* with a random outcome, we understand the complex conditions under which a particular phenomenon occurs.

Example 2.1

Imagine that a box contains black and white balls. One ball is taken out of the box (this will be the experiment with a random outcome). Then the event $A = \{\text{the extracted white ball}\}$, and the event $B = \{\text{the extracted black ball}\}$.

Events are called *compatible* in this experiment, if events can occur at the same time.

https://doi.org/10.1016/B978-0-12-822066-5.00002-9

Example 2.2

Let black and white balls be in one box, and yellow and green in a second. One ball is taken out of each box. Event $A = \{$a white ball was extracted from box No. 1$\}$, and event $B = \{$a yellow ball was extracted from box No. 2$\}$. These two events can occur simultaneously, which means they are compatible.

Two events are called *incompatible* in a given experiment if they cannot occur simultaneously. For example, the extraction of a black ball and the extraction of a white ball from one box in one experiment are incompatible events, since at a time according to the conditions of the experiment you can get only one ball (either black or white).

A set of events is *jointly* or *collectively exhaustive* if they are pair-wise incompatible—that is, no two of them can happen simultaneously but at least one of the events must happen.

Opposite (or *complement*) events are a collectively exhaustive set of two events. They are denoted by a letter and a letter with a dash, for example, event A and the opposite event not A, which is denoted as $\bar{A}$.

Events are called *equally likely events*, if there is no reason to believe that one of them occurs more often than others. Each equally possible event that can occur in a given experiment is called an elementary outcome. The elementary outcomes at which an event occurs are called the *elementary outcomes favorable to this event.*

In this case, *the probability* of the event A is considered to be the ratio of the number m of outcomes favorable to event A to the number n of all possible outcomes of a given experiment:

$$P(A) = m/n. \tag{2.1}$$

So, probability is a numerical description of the likelihood of an event or the likelihood that an assumption is true. In general, a value of probability of 1 corresponds to certainty. An impossible event has a probability of 0. If the probability of an event occurring is $P(A)$, then the probability of its nonoccurrence is $1 - P(A)$. In particular, a probability of $1/2$ means an equal probability of occurrence and nonoccurrence of an event. As we see, the probability concept is associated with the randomness of an event.

The empirical definition of probability is associated with the frequency of the occurrence of an event, based on the fact that, with a sufficiently large number of tests, the frequency should tend to an objective degree of possibility of this event (Chung, 2014; Editor(s): Terrence L. Fine, Theories of Probability, 1973).

For example, the initial value for quantifying anthropogenic risk is the probability of an accident. The probability of the implementation of the considered initial emergency for the year (frequency of the initial event) was traditionally determined on the basis of normative and reference documentation. Assessment of the expected frequency of accidents at a facility is also possible based on the analysis of statistics on accidents at similar facilities. This requires the application of statistical methods. As we know, mathematical statistics is a science that develops mathematical methods for systemizing and using statistical data for scientific and practical conclusions.

In many of its functions, mathematical statistics is based on probability theory, which makes it possible to assess the reliability and accuracy of conclusions made on the basis of limited statistical material (for example, to estimate the necessary sample size to obtain the results of the required accuracy in a sample survey) (Ash & Doleans-Dade, 1999; Pugachev, 1984).

Statistical analysis should identify significant patterns in the frequency and spectrum of accidents and use these patterns in practice to calculate the frequency of accidents of the investigated object. To obtain the most reliable estimate of the frequency of initiating events, it is necessary that the statistics on accidents that have already occurred allow us to identify the influence of significant factors (natural, man-made, technical) characteristic of this object on the frequency of possible accidents, i.e., they need to be representative (Pugachev, 1984).

Next, we consider the main actions with probabilities.

The *sum* of $A + B$ of events A and B is the event consisting of the occurrence of at least one of them, i.e., only event A, or only event B, or events A and B at the same time. The probability of the sum of incompatible events is equal to the sum of their probabilities:

$$P(A+B) = P(A) + P(B). \tag{2.2}$$

The *product* of events $A \times B$ of events A and B is an event consisting of the occurrence of events A and B simultaneously.

Two events are called *dependent* if the probability of one of them depends on the occurrence or nonoccurrence of the other. Two events are called *independent* if the probability of occurrence of one of them does not depend on the occurrence or nonoccurrence of the other.

Example 2.3

Let two balls be sequentially removed from a box containing 6 white and 10 black balls (and not returned to the box). Event $A = \{$the first ball extracted from the box is black$\}$ and the event $B = \{$the second ball extracted from the box is black$\}$. Find out if events A and B are dependent. Suppose that the event A has occurred, i.e., the first ball extracted is black. After that, 15 balls remain in the box, of which 9 are black. Therefore, the probability of the event is B $P(B) = 9/15$, since a total of 15 options are possible, 9 of which favor event B. Now suppose that the event A did not happen, but the opposite event occurred; that is, $\bar{A} = \{$the first ball extracted from the box is not black$\} = \{$white ball is extracted from the box first$\}$. Then the box contains 19 balls, of which 10 are black. Then the probability of the event is B $P(B) = 10/15$, since a total of 15 options are possible, 10 of which favor event B.

As follows from the analysis, the probability of the occurrence of the event B depends on the occurrence or nonoccurrence of event A. Thus, these two events are dependent.

The probability of the product of two independent events is equal to the product of their probabilities:

$$P(AB) = P(A)P(B), \tag{2.3}$$

where A and B are independent events.

Example 2.4

Consider a system consisting of two devices working independently from each other. The system functions normally only with the simultaneous operation of these two devices. The probabilities of proper operation of the first and second devices are 0.9 and 0.95, respectively. It is required to find the probability of the normal functioning of the entire system.

Event $A = \{$the first device is running$\}$, event $B = \{$the second device is running$\}$. By the condition of the problem, $P(A) = 0.9$, $P(B) = 0.95$. The product of events $AB = \{$both devices work$\} = \{$the whole system works$\}$. Since events A and B are independent (according to the conditions of the task, the devices operate independently for each other), $P(AB) = P(A)\ P(B) = 0.9 \times 0.95 = 0.855$.

The conditional probability $P(B|A)$ is the probability of event B, provided that event A has occurred. Then for dependent events:

$$P(B|A) \neq P\left(B|\overline{A}\right). \tag{2.4}$$

On the contrary, for independent events:

$$P(B|A) = P\left(B|\overline{A}\right) = P(B). \tag{2.5}$$

Example 2.5

If we consider Example 2.3 for conditional probability, then $P(B \mid A)$ is probability of the black ball is the second to be extracted from the box, provided that the black ball is also extracted first. It is defined as probability of the second ball to be extracted from the box, provided that the black ball was first extracted from the box. It will be equal to 9/15.

$P(B \mid \bar{A})$ is probability of the black ball is the second to be extracted from the box, provided that the black ball is not the first to be pulled. The probability is equal to 10/15.

Since the probabilities obtained are not equal, the events are dependent.

For dependent events A and B, the following statement is true:

$$P(AB) = P(A)P(B|A). \tag{2.6}$$

Example 2.6

Let us return to the consideration of Examples 2.3 and 2.5. We find the probability that two black balls are extracted from the box, that is, we find the probability of the event AB. Probability P (two black balls sequentially extracted from the box) = $P(AB) = P(A)P(B \mid A) = P$ (the first black ball is extracted from the box) $\times$ P (the second ball is extracted from the box black, provided that a black ball is also extracted) = $10/16 \times 9/15 = 3/8$.

The probability of the sum of two compatible events A and B is equal to the sum of their probabilities minus the probability of the product of these events:

$$P(A + B) = P(A) + P(B) - P(AB). \tag{2.7}$$

If events A and B are incompatible, then AB is an impossible event, that is, $P(AB) = 0$, and formula (2.6) turns into formula (2.2).

Next, we consider the process of finding the so-called full probability, when it is necessary to determine the probability of an event that can occur with one of the incompatible events to form a complete group of events. Such a task may arise, for example, during the analysis of an emergency, consisting of chains of possible options for the development of the situation and the need to separate the scenarios. The simplest and most visual approach is to build a probability tree.

When constructing a probability tree, experiments are usually represented graphically in the form of circles. Each outcome is a tree branch and is indicated by a solid line emerging from the corresponding circle (experience). Near each line, the probability of the corresponding outcome is indicated. The sum of the probabilities on the branches emanating from one experience (circle) is equal to one. Moving along the branches and successively multiplying the probabilities, at the end of the path we get the probability of a complex event for each scenario. Having added the necessary probabilities, we will find the probability of the desired event.

Example 2.7

We construct a simplified tree of event probabilities during the destruction of a vessel under pressure, in which the possibility of internal explosions is excluded (Fig. 2.1).

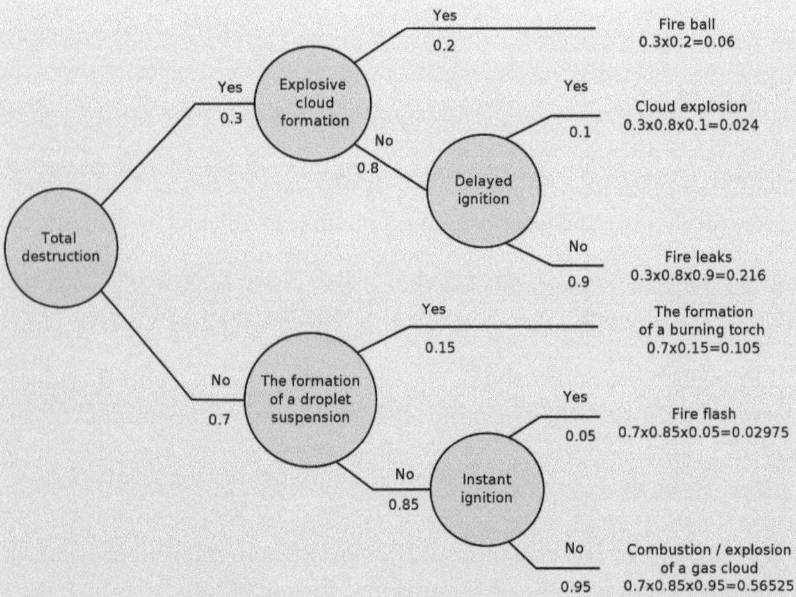

Fig. 2.1 The probability tree of events during the destruction of a vessel under pressure.

In this case—the destruction (partial or full) of a vessel with a dangerous substance (initial event)—events are possible such as: release into the environment of a hazardous substance; formation and spread of a strait of a hazardous substance and its partial evaporation; the formation of an explosive concentration of vapors of a hazardous substance in the air; ignition of vapors of a hazardous substance and/or spill of a hazardous substance in the presence of an ignition source; combustion of the fuel-air mixture; spill fire; people, equipment, and/or environmental objects getting into the zone of possible damaging factors; and subsequent development of the accident if the equipment involved contains hazardous substances.

The numerical values of the probabilities, of course, may differ from the statistical data for real cases and are used here solely for clarity of calculations. For example, we can see that the most likely resultant event with these initial data is a gas cloud explosion (with a probability of 0.56525), and the least likely is the formation of a fireball (with a probability of 0.06).

In conclusion of this section, we consider another concept of probability theory, which is widely used in risk analysis theory for interrelated events. Bayes formula is fundamental in the elementary theory of probability. It makes it possible to determine the probability of an event, provided that another event, interdependent with it, has occurred. In other words, according to the Bayes formula, the probability can be more accurately calculated taking into account both previously known (a priori) information and the data of new observations:

$$P(A|B) = \frac{P(B|A)P(A)}{P(B)}, \tag{2.8}$$

where A and B are events and $P(B) \neq 0$; $P(A)$ is the a priori probability of hypothesis A; $P(A|B)$ is probability of hypothesis A upon occurrence of event B (posterior probability); $P(B|A)$ is probability of occurrence of event B with the hypothesis A true; and $P(B)$ is the total probability of event B.

Example 2.8

The manufacturer claims that the probability of facility failure is 5% (hypothesis A, a priori probability). It is likely that the facility rejected by the manufacturer is in working condition (15%), as well as the likelihood that the a priori operational facility will fail (5%). It is necessary to determine the probability of the event B that the facility will indeed fail during operation. We consider two consecutive experiments: the type of the object (a priori failure or not failure) and the test of the object. Let's construct a tree of probabilities (Fig. 2.2).

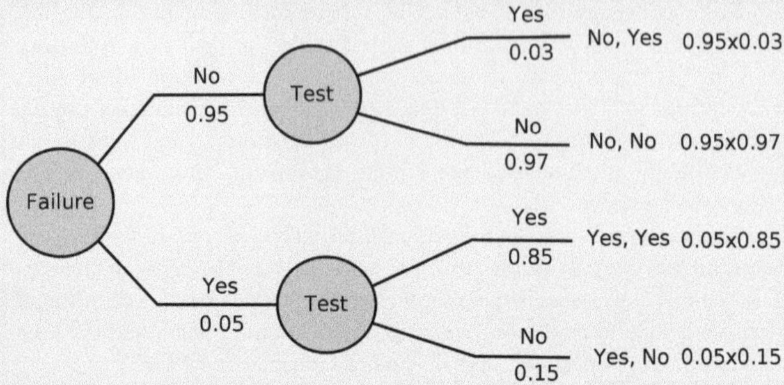

Fig. 2.2 The probability tree for determining Bayes probability (Example 2.8).

The probability that the facility will fail during the tests is $0.97 \times 0.03 + 0.05 \times 0.85 \approx 0.0672$. The probability that a priori an inoperative facility fails is $0.05 \times 0.85 = 0.0425$. Then the desired probability according to formula (2.8) will be $0.0425/0.0672 \approx 0.632$.

The peculiarity of the Bayes formula is that for its practical application a large amount of data and calculations are required; therefore, Bayesian estimates began to be actively used only after the introduction of new computer and network technologies, including big data technologies.

2.1.2 Risk definition and risks calculation

In the most general sense, In accordance with ISO 31000:2018, *risk* includes the effects of any of the forms of uncertainty on objectives. Risk is also commonly defined as a combination of the probability and consequences of adverse events. This definition is very general and can be applied to any industry or organization. In the field of process safety, risk is more associated with process industry, emergencies, chemical and physical processes, toxic hazards, fires, explosions, etc. In this sense, risk can be defined as the probability of a hazardous factor getting out of control and the severity of the consequences, expressed by the degree of manifestation. The risk assessment includes identifying risks, analyzing risks, and evaluating risk. In addition, the following terms and definitions are needed to understand the process of risk assessment.

A source of risk is an element that can lead to a risk. An event is the occurrence or change of a specific set of circumstances. An event can be a source of risk. The consequence is the result of an event affecting the goal. The likelihood is the chance that something happens, is defined, measured or determined objectively or subjectively, qualitatively or quantitatively, and described using general terms or mathematically (for example, probability or frequency over a certain period of time). Probability is a measure of the chance of occurrence, expressed as a number from 0 to 1, where 0 is impossibility, and 1 is absolute certainty in accordance with IEC 31010:2019. Control is a measure that supports and (or) modifies risk (IEC 31010:2019, Risk management—Risk assessment techniques, 2019).

Risk analysis is the procedure helping to understand the nature of the risk and its features, including, when necessary, the level of risk. It also allows us to consider the specifics of the development of various types of accidents, as well as the determination of quantitative indicators of the associated social, material, and environmental damage. Risk analysis includes a detailed review of uncertainties, sources of risk, consequences, probability, events, scenarios, and controls and their effectiveness. An event can have various causes and consequences and can affect different goals.

The combination of consequences and probability (usually as their product) forms a risk category as a quantitative criterion for assessing safety, making it possible to obtain a universal scale for comparing hazards of various origins (ISO 31000:2018).

Typically, the risk of an accident is calculated in units of time-related damage. The determining ratio for predicting risk levels can be represented as the sum of the products of the frequency of occurrence of hazardous events by the amount of the corresponding damage.

Summation is performed over the entire set of emergency processes that may occur at the facility (Guidelines for Safe Storage and Handling of High Toxic Hazard Materials, 1988). From the above ratio, it follows that the forecast of the level of emergency danger is associated with a frequency analysis of possible emergency processes and with the forecast of damage in potential accidents. Unlike other approaches to assessing process safety, the risk methodology, within the framework of a system analysis, makes it possible to:

- investigate the cause-effect mechanisms of various accidents and predict their frequency;
- take into account the impact of technological, meteorological, regional, and a number of other features on the nature and extent of the consequences of accidents; and
- optimize management decisions to improve the safety of the facility in conditions of limited resources.

In other words, the risk methodology, together with the application of the big data concept, makes it possible to implement the "anticipate and warn" principle instead of the existing "react and correct" principle.

Chemical process quantitative risk analysis (CPQRA) is an interdisciplinary science, which is based on mathematical modeling of the behavior of a dangerous substance that has entered the environment, including toxic damage to biosphere objects, fires, explosions, and their consequences (Guidelines for Chemical Process Quantitative Risk Analysis, 1989). It uses probabilistic methods of reliability theory, hydraulics, physical and analytical chemistry, applied meteorology, toxicology, engineering psychology, etc. This methodology focuses on the chemical, petrochemical, and refining industries. Some features of these industries include:

- the use of various chemicals;
- possible chemical reactions;
- lack of process standardization;
- materials may not have well-defined properties;
- a big difference between such parameters as the type and age of the installation, the location of the surrounding population, degree of automation, and type of equipment; and
- usually simultaneous impact of such factors as fire, explosion, toxicity, and environmental pollution.

The CPQRA method considers all possible cases or accidents due to some activity and estimates the probability or frequency of each event and its related consequences through a number of accidents. The information obtained can then be summarized and presented in a convenient form. The classical scheme of quantitative risk assessment is presented in Fig. 2.3.

Thus, CPQRA risk is a statistical assessment category, which is a vector multicomponent function.

(*Guidelines for Hazard Evaluation Procedures. 2nd Edition with Worked Examples*, 1992; Guidelines for Technical Process Safety, 1987):

$$Risk = F(S, C, F), \tag{2.9}$$

where S is a hypothetical scenario, C is estimated consequence, and F is estimated frequency.

This function can be extremely complex, and there can be many numerically different risk measures (using different risk functions) calculated from a given set of S, C, F. Big data can be used both to describe scenarios S and to determine the frequency of the occurrence of a hazard F.

2.1.3 Problems of data acquisition for risks calculation

The generally accepted measures of the level of hazard, as already described above, are risk assessments. They allow a quantitative analysis of the level of

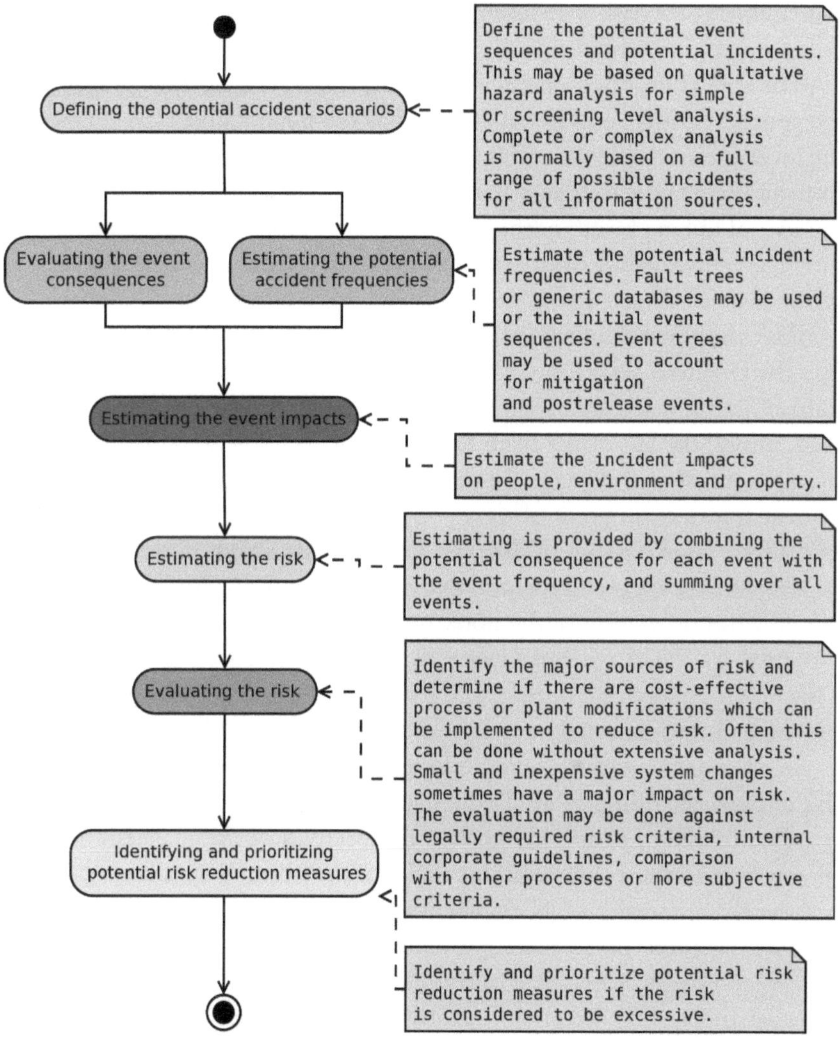

Fig. 2.3 CPQRA flow diagram.

hazard relative to specific risk recipients. Analysis of risk assessments makes it possible to differentiate hazardous industrial objects primarily by the threat that they pose to humans and the natural environment, and makes it possible to differentiate territories according to the level of potential danger. In terms of risk assessments, safety criteria are expressed.

A variety of hazard manifestations corresponds to a variety of assessments, depending on the operating mode of the research object and risk assessment associated with the consequences of an accident at the facility. The latter are called emergency risk assessments. These two types of risk are sometimes called real and potential.

The allocation of emergency risk assessments to a separate category, in the general case, is conditional and reflects the quantitative side. Typically, the level of emergency hazard is significantly higher than the level of hazard from an object operating in a normal mode, when the expected impacts on the state of human health or the state of the environment are insignificant. In this regard, emergency risk assessments, as a rule, characterize the upper limit of the level of danger generated by an industrial facility.

Risk assessments can be classified based on who or what is at hazard, that is, on a risk recipient. So we can distinguish risk assessments relative to the state of human health, risk assessments relative to the state of the environment, etc. And, finally, the last of the main features by which risk assessments are classified is a measure of damage. Thus, risk assessment is a very complex process from an information point of view. To obtain risk assessments, it is necessary to collect a large amount of data and build models based on these data.

The chemical process quantitative risk analysis (CPQRA) mentioned in the previous subsection of the book is a part (or a step) of the methodology proposed by the American Center for Chemical Process Safety (CCPS). This methodology combines risk calculation, hazard analysis quality techniques, and decision-making methods for the safety of the facility.

In accordance with the methodology, it is recommended to conduct a risk analysis according to a scheme that includes the following main steps:

- determination of specific goals and objectives of the analysis;
- analysis of the technological specifics of the object with a description of the characteristics of its environment;
- identification of hazards, possible accidents, and scenarios of their development;
- assessment of the frequency (probability) of accidents and the likelihood of the implementation of characteristic scenarios of their development;
- assessment of consequences (that is, values of the characteristics of the damaging factors and measures of negative impact on potential recipients) using models for calculating physical processes and effects that occur during the implementation of various accident scenarios;
- assessment of the actual risk through the "combination" of the consequences and probabilities of the implementation of all possible accident scenarios and the construction of risk fields; and
- risk management, which consists of developing an optimal strategy for ensuring human safety and environmental protection.

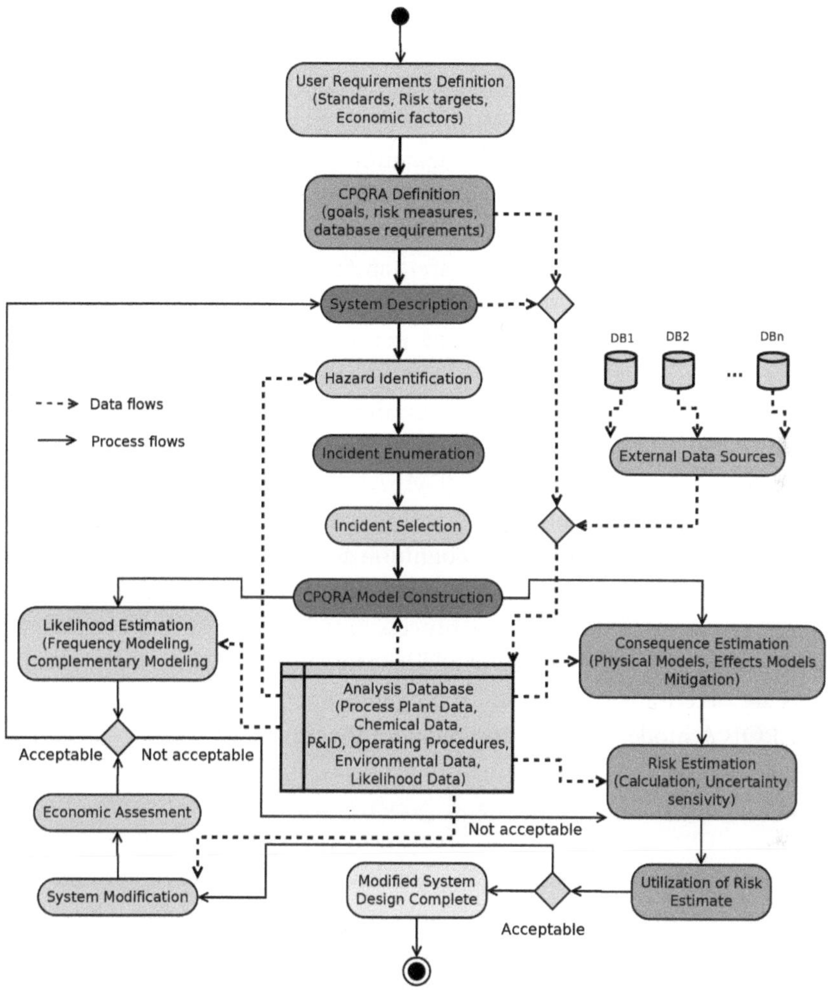

Fig. 2.4 Process and data flows in the framework of CPQRA methodology

A short description of each of the steps is presented in Fig. 2.4 (Guidelines for Chemical Process Quantitative Risk Analysis, 1989).

The CPQRA definition transforms user requirements into assessment objectives. Risk measures are necessary to determine the final scope of the CPQRA study. Further coverage of the study is chosen having in mind the specified objectives and the available resources; this may include, for instance, the evaluation of affected information systems and data flows failures. For this, specific requirements for information (required item, data formats, accessibility, etc.) and hardware and software features for the construction of the analysis data base are considered.

System description is the integral data set of the process/plant information required for the risk calculation. It usually consists of process flow diagrams (PFD), piping and instrumentation diagrams (P&ID), drawings, operating and maintenance procedures, technology documentation, process chemistry, the site location, the environment, weather data, and thermophysical property data. These data are transferred into the analysis database for use in CPQRA.

Hazard Identification is the next step of CPQRA. This is important because a missed hazard is a hazard that is not being analyzed. This step includes experience, engineering codes, and checklists analysis, detailed process knowledge, equipment failure experience, hazard index techniques, failure modes and effects analysis (FMEA), "what-if" analysis, hazard and operability (HAZOP) studies, and preliminary hazard analysis (PHA).

Incident enumeration is the recognition and arranging of all known incidents. It is also critical to take all the possible incidents into account.

The selection of incidents is a process in which one or more significant incidents are selected to represent all identified incidents, the results of incidents are determined, and cases of incident outcomes are developed.

CPQRA model construction includes the choice of appropriate consequence models, likelihood estimation methods, and their integration into an overall algorithm to produce and present risk estimates for the studied system.

Consequence estimation is the methodology used to determine the potential outcomes for damage or injury from specific incidents. These outcomes are analyzed using source and dispersion models and explosion and fire models. Effects models are then used to determine the consequences to people or structures. Evasive actions such as sheltering or evacuation can reduce the magnitude of the consequences and these may be included in the analysis.

Likelihood estimation is the methodology used to estimate the frequency or probability of occurrence of an incident. Estimates may be obtained from historical incident data on failure frequencies, or from failure sequence models, such as fault trees and event trees.

Risk estimation combines the consequences and likelihood of all incident outcomes from all selected incidents to provide one or more measures of risk. It is possible to estimate a number of different risk measures from a given set of incident frequency and consequence data, and an understanding of these measures is provided. The risks of all selected incidents are individually estimated and summed to give an overall measure of risk.

Utilization of risk estimates is the process by which the results from a risk analysis are used to make decisions, either through relative ranking of risk reduction strategies or through comparison with specific risk targets.

The last CPQBA step (utilization of risk estimates) is the key step in a risk assessment. It requires the user to develop risk guidelines and to compare the risk estimate from the CPQRA with them to decide whether further risk reduction measures are necessary. This step has been included as a CPQRA component technique to emphasize its overall influence in designing the CPQRA methodology. Guidelines for decision analysis are contained in Tools for Making Acute Risk Decisions (Tools for Making Acute Risk Decisions, 1992).

The development of the analysis database is a critical step in all the CPQRA. Thus, the problem of data acquisition for risk calculation is the cornerstone of the CPQRA methodology. In addition to the data from the system description, this database contains various kinds of environmental data (e.g., land use and topography, population and demography, meteorological data) and likelihood data (e.g., historical incident data, reliability data) needed for the specific CPQRA. Much of this information must be collected from external (outside company) sources and converted into formats useful for the CPQRA. The need to use external databases can lead to additional difficulties associated with permission to access data, data transfer, information security, etc. Here are a few examples of the databases on the listed topics. These examples are provided only to give an idea of the possible breadth and heterogeneity of the required data.

The Topographic Database is a dataset depicting the terrain of all of Finland. The key objects in the Topographic Database are the road network, buildings and constructions, administrative borders, geographic names, land use, waterways, and elevation. Aerial photographs, scanning data, and data provided by other data providers are utilized in updating the Topographic Database. The updating is done in close cooperation with the municipalities. Field checks in the terrain are also needed to some extent, mostly as regards the classification of features (Topographic Database, n.d.).

The International Network for the Demographic Evaluation of Populations and Their Health (INDEPTH) is a global network of research centers that conduct longitudinal health and demographic evaluation of populations in low- and middle-income countries (LMICs). INDEPTH aims to strengthen global capacity for Health and Demographic Surveillance

Systems (HDSSs), and to mount multisite research to guide health priorities and policies in LMICs, based on up-to-date scientific evidence. The data collected by the INDEPTH network members constitute a valuable resource of population and health data for LMIC countries. The INDEPTH Data Repository aims to make well-documented, anonymized, longitudinal microdata from these Centers available to data users. As of October 13, 2016, the library contains 66 surveys, 3181 citations, and 1396 variables (The INDEPTH Data Repository, n.d.).

The National Weather Service (NWS) provides weather, water, and climate data, forecasts, and warnings for the protection of life and property and enhancement of the national economy (The National Weather Service (NWS), n.d.).

2.1.4 Big data and risk calculation

When calculating and predicting the process safety risks, big data can be used for dynamic risk assessment and automatic report generation. This enables hazardous production personnel to identify safety threats at an early stage, taking the necessary preventive measures to address them. In this way, a number of incidents involving the sudden shutdown of industrial plants or accidents can be avoided.

This problem is especially relevant for oil refineries, which are hazardous facilities. Modern oil cracking plants are recognized as some of the most efficient in the refining industry. They are equipped with automatic and automated control and monitoring systems based on intelligent software and a distributed sensor system. Such systems generate hundreds of thousands of data points. However, the magnitude of the risks and reliability indicators associated with the rise of the regenerated catalyst (as well as their change over time) remain unknown (Pariyani, Oktem, & Grubbe, n.d.; Resolve Your Process Risks at Their Initiation Stage, n.d.). This fact creates difficulties in the management of the facility, reduces the efficiency of the cracking process, and increases the likelihood of industrial accidents. The problem is also exacerbated due to aging equipment and a shortage of experienced operators who own knowledge of not only the subject area, but also the basics of information technology.

Thus, over the past decade, production facilities have generated a huge amount of data, but not all data has become useful information.

On average, more than 5 billion data points from 320 sensors that measure every second are recorded at a chemical plant every 6 months. The information hidden in big data is an indicator that can help enterprises assess dynamically changing risks and avoid the financial losses of tens of billions of dollars that the chemical and petrochemical industry suffers annually due to unexpected production shutdowns and accidents. Studies show that the use of big data technologies based on information processing directly using advanced methods of data mining reveals hidden relationships and trends in risk assessment that were previously inaccessible. This innovative approach to predictive risk assessment can help enterprises prevent accidents and unexpected outages (Protecting Refractory Linings in FCC Units, n.d.; Niccum, 2011).

Improved process risk management techniques have been the main result of the widespread adoption of the process safety management (PSM) standard. The PSM standard was developed by the US Occupational Safety and Health Administration (OSHA) (Process safety management of highly hazardous chemicals, n.d.) to maintain and improve the safety, health, and productivity of manufacturing operations. Over the past decade, progress has been made in assessing the risks associated with the use of new information technologies, including big data technologies. Let us consider briefly modern methods of calculating and analyzing risks from the point of view of the prospects for implementing big data.

Many enterprises conduct *safety, health, and environmental audits* using internal teams and third-party consulting companies, which incur significant costs. The frequency and effectiveness of internal safety audits (usually once a year or less) largely depend on the capabilities of a particular enterprise. In most cases, safety experts, with some support from engineers, operators, and managers, periodically review work procedures and safety records, and conduct a limited number of safety interviews. An integral part of these audits is the analysis of incident history and observed errors reported by employees. The latter depends on the safety culture at the enterprise and cannot always give a real picture of risks.

Operations management and production intelligence tools (Sower, Jaideep, & Savoie, 1997) are based on key performance indicators (KPIs). Moreover, the information system provides monitoring of the performance of operations and assessment of the availability and effectiveness of equipment. This technique is focused on trend analysis, reporting, and visual analytics of the selected time slice of data. This helps employees to track the variability of various parameters over a period of time (shift, day, week, etc.). However, this approach is not effective in the analysis of big data, especially when it comes to identifying anomalies and potential sources of risk. This requires a comparison of working conditions with their normal to detect new changes, which is not part of the risk assessment method under consideration.

Condition monitoring tools (Xuemei, Shaobo, & Ansi, 2018) detect abnormal situations in real or near real time, comparing the current parameters of the installation with the expected (reference) behavior and warning employees about a dangerous deviation. There are model-driven (based on quantitative process models) and data-driven tools (based on clustering and dimensional reduction approaches) that help operators take immediate corrective actions when they receive alerts. In this way, they can signal a hazard more effectively than traditional alarms with fixed thresholds.

However, since these tools are designed to monitor the process in real time, they do not focus on the history of the development of risks and the likelihood of incidents over a certain period of time (days, weeks, months). This feature is a significant drawback in assessing the magnitude of risks and process characteristics, which is important information for enterprise managers, engineers, and reliability specialists to make strategic decisions.

Quantitative risk assessment (QRA) has been discussed in some detail in the previous paragraphs. As a rule, QRAs are conducted once every few years by most enterprises. They use various data sources available to the industry, such as incident data, material safety data, and equipment and human safety data, to identify incident scenarios and assess risks by identifying the likelihood of failure and its potential consequences. They help users identify possible ways to reduce risk. Since the QRA mainly includes incident and fault data (with the exception of routine process and alarm data that provides information on the possible causes of incidents), it has limited predictive power.

In contrast to the considered methods, the early risk detection technique proposed by *near-miss management* allows industrial enterprises to identify and reduce process risks by receiving early warning signals of risk development and taking preventive measures. *Dynamic risk analysis* [8] (Shenae, Gabriele, Genserik, & Nicola, 2019) is an advanced methodology for risk prevention and detection that implements a distinctive approach of identifying hidden near-misses in the process data. Based on large-scale machine learning and risk management, dynamic risk analysis technology scans the entire spectrum of process data and identifies critical points called hidden accidents that are the forerunners of potential failures. This risk development information is available to all levels of the enterprise's personnel, including managers,

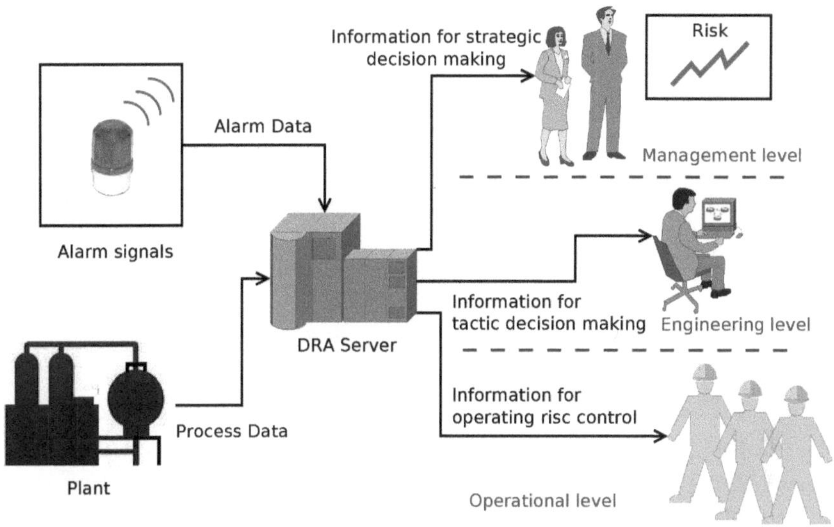

Fig. 2.5 The generalized information structure of dynamic risk analysis (DRA).

engineers, supervisors, operators, and maintenance teams. The result of the implementation of the system is the observability of production processes, active risk management, and effective interaction of employees at all levels (Fig. 2.5).

Example—Oil refinery shutdown. It is known that the riser of the regenerated catalyst is one of the critical elements of a catalytic cracking unit. A sudden failure of this element from time to time leads to malfunctions in the installation and shutdowns (Fig. 2.6).

The dynamic risk analysis study used a risk analysis methodology to analyze data quickly for more than 18 months. It was found that the deterioration of the situation occurred gradually over several months. Installed standard factory monitoring systems and personnel could not see these changes. Rising levels of risk were identified for several variables (long before a blackout), indicating significant deviations from normal behavior, invisible in the standard construction of graphs, trend lines, and data visualization. In particular, a key differential pressure variable was determined (for one of the pressure pipes of the reduced catalyst), which had the greatest impact on indicators of the possible shutdown. One of the leading indicators (probability of disconnection) gradually increased to 50% over several months. Available forward-looking information helps staff monitor the effectiveness of existing risk reduction measures and identify problems in

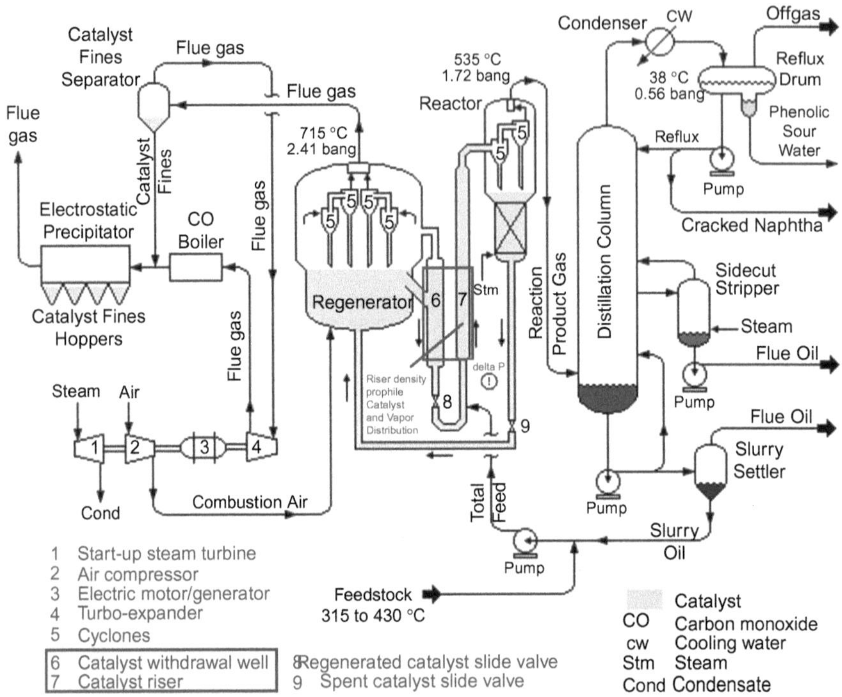

Fig. 2.6 Schematic flow diagram of a fluid catalytic cracker.

the early stages, so that management can strategically direct the necessary resources to the most critical production areas in a timely manner.

2.2 Standards for safety

The main functions of the standards are traditionally to increase the level of safety regarding the lives and health of citizens, property of individuals and legal entities, state and municipal property, facilities taking into account the risk of emergency situations of natural and man-made nature, increase the level of environmental safety, safety of animals and plants, ensuring the competitiveness and quality of products (works, services), the uniformity of measurements, the rational use of resources, the interchangeability of technical means (machines and equipment, their components, components, and materials), technical and information compatibility, the comparability of the results of research (testing) and measurements, technical and economic-statistical data, analysis of the characteristics of products (works, services), execution of government orders, voluntary confirmation of conformity of products (works, services), and facilitating compliance with technical regulations. Standards play a very important role in ensuring process safety. The norms, standards, and rules of process safety establish mandatory requirements and give recommendations on ensuring the safety of technological processes at hazardous production facilities, including with respect to the procedure for action in the event of an accident or incident at a hazardous production facility.

2.2.1 OSHA

The Occupational Safety and Health Administration (OSHA) was created by the U.S. government to ensure safe and healthful working conditions for working people by setting and enforcing standards and by providing training, outreach, education, and assistance (Occupational Safety and Health Administration, n.d.). The standards cover the following branches, including process safety:

- general industry;
- construction;
- maritime;
- agriculture;
- recordkeeping;
- whistleblower; and
- preambles to final rules, etc.

The standards provided by OSHA have hierarchical structure related to the industries, main processes, and safety fields.

One of the key documents for chemical process safety is OSHA 1910.119 "Process safety management of highly hazardous chemicals" (Fig. 2.7) (Process safety management of highly hazardous chemicals, n. d.). This subpart of the part 1910 "Occupational Safety and Health Standards" (1910 Subpart H titled "Hazardous Materials") consists of requirements for preventing or minimizing the consequences of catastrophic releases of toxic, reactive, flammable, or explosive chemicals. These releases may result in toxic, fire, or explosion hazards. Section 1910.119(d), named "Process safety information," requires the employer to complete a compilation of written process safety information before conducting any process hazard analysis required by the standard in accordance with the set schedule.

Process safety information can be considered a very important part of process safety management (PSM). It provides personnel with knowledge of how to act to prevent hazardous situations, taking into account production equipment and processes (*Guidelines for process safety documentation*, 1995).

The collected process safety information enables the employer and the employees involved in PSM to identify and analyze the hazards posed by the processes involving highly hazardous chemicals. This process safety information includes information pertaining to the hazards of the highly hazardous chemicals used or produced by the process, information pertaining to the technology of the process, and information pertaining to the equipment in the process.

This information structure shows the variety and large volume of data it is necessary to analyze to ensure process safety. It is necessary at every stage of PSM. These information flows are indicated as $I_1 - I_{12}$ in Fig. 2.7.

The three main sections of data are described and listed below.

Section 1910.119(d)(1) Hazards of the Chemicals Used in the Process. Accurate written and complete information on the chemicals used in the process, process equipment, and process technology is fundamental for sufficient process safety management and process hazard analysis. The assembled data are a basic resource for users including the team providing the process hazard analysis (1910.119 (e), information flow I_2), those developing operating procedures (1910.119 (f), information flow I_3) and training programs (1910.119 (g), information flow I_4), contractors (1910.119 (h), information flow I_5) whose employees will be working with the process (1910.119 (c),

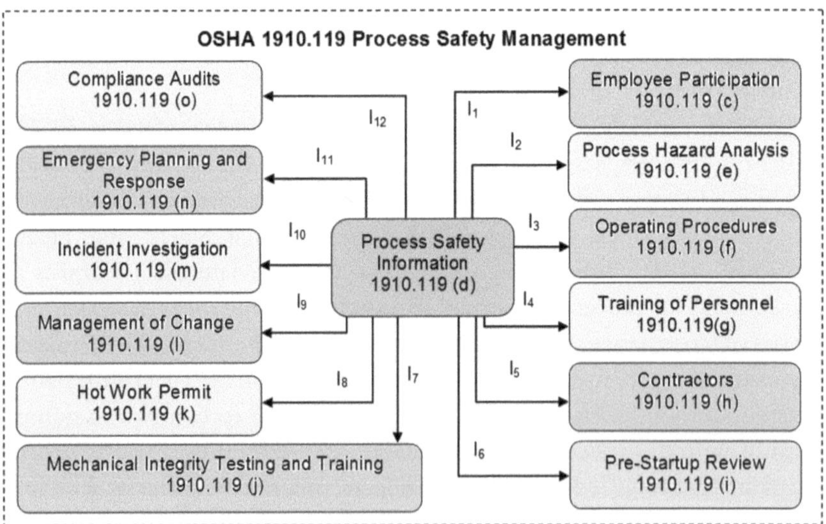

Fig. 2.7 The structure of process safety management (OSHA 1910.119).

information flow I_1), those providing prestartup reviews (1910.119 (i), information flow I_6), together with local emergency preparedness planners (1910.119 (n), information flow I_{11}), enforcement officials, and insurers (1910.119 (j), information flow I_7).

The information collected on chemicals, including intermediates, should provide sufficient coverage to assess accurately fire and explosion characteristics, reactivity hazards, health and safety hazards for workers, and the effects of corrosion and erosion on process equipment and monitoring tools. Current material safety data sheet (MSDS) data can be used to help with this, but should be supplemented with process chemistry information, including runaway reaction and the risk of overpressure, if applicable.

Thus, section *1910.119(d)(1) Hazards of the Chemicals Used in the Process* consists of the following data:

- 1910.119(d)(1)(i)—Toxicity information;
- 1910.119(d)(1)(ii)—Permissible exposure limits;
- 1910.119(d)(1)(iii)—Physical data;
- 1910.119(d)(1)(iv)—Reactivity data;
- 1910.119(d)(1)(v)—Corrosivity data;
- 1910.119(d)(1)(vi)—Thermal and chemical stability data; and
- 1910.119(d)(1)(vii)—Hazardous effects of inadvertent mixing of different materials that could foreseeably occur.

Section 1910.119(d)(2)—Information pertaining to the technology of the process. Information about the process technology is part of the process safety information package and should include criteria established by the employer for the maximum inventory levels for the process chemicals; the limits beyond which the violated conditions will be considered; and a qualitative assessment of the consequences or results of deviations that may occur when working outside the established process limits (1910.119 (k), (l), (m), (o), information flow I_8, I_9, I_{10} I_{12}). Employers are encouraged to use charts to help users understand the process.

The flowchart is a simplified diagram used to display the main processing equipment and interconnected production lines and flow rates, flow composition, temperatures and pressures, when necessary for basic understanding a technological process.

Process flow diagrams are more complex and show all the main flow streams, including valves, to improve understanding of the process, as well as pressure and temperature on all supply and product lines in all main vessels, as well as to and from collectors and heat exchangers, as well as control points pressure and temperature. In addition, information on structural materials, pump performance and head, compressor power, as well as design pressure and vessel temperatures are displayed if necessary. Furthermore, process flow charts typically show the main components of control loops along with key utilities.

The section *1910.119(d)(2) Information concerning the technology of the process* must include at least the following:

- 1910.119(d)(2)(i)(A)—A block flow diagram or simplified process flow diagram (see Appendix B to this section);
- 1910.119(d)(2)(i)(B)—Process chemistry;
- 1910.119(d)(2)(i)(C)—Maximum intended inventory;
- 1910.119(d)(2)(i)(D)—Safe upper and lower limits for such items as temperatures, pressures, flows, or compositions;
- 1910.119(d)(2)(i)(E)—An evaluation of the consequences of deviations, including those affecting the safety and health of employees; and
- 1910.119(d)(2)(ii)—Where the original technical information no longer exists, such information may be developed in conjunction with the process hazard analysis in sufficient detail to support the analysis.

Section 1910.119(d)(3)—Information pertaining to the equipment in the process. These data are necessary at the PSM parts of process hazard analysis (1910.119 (e), information flow I_2), operating procedures (1910.119 (f), information flow I_3), training programs; (1910.119 (g), information flow I_4), contractors procedures (1910.119 (h), information flow I_5), employees participation (1910.119 (c), information flow I_1), prestartup reviews (1910.119 (i), information flow I_6), local emergency preparedness planning (1910.119 (n), information flow I_{11}), and insurance (1910.119 (j), information flow I_7).

Piping and Instrument Diagrams (P&IDs) may be a more suitable type of diagram to show some of the above details, as well as display information for the piping designer and engineering staff. P&IDs should be used to describe the relationships between equipment and instrumentation, as well as other relevant information that will enhance clarity. Computer software that use P&ID or other schemes useful for the information package can be used to satisfy this requirement. An example of P&ID is shown in Fig. 2.8.

The information pertaining to process equipment design must be documented is accordance with certain codes and standards.

The section *1910.119(d)(3)(i)—Information pertaining to the equipment in the process* shall include:

- 1910.119(d)(3)(i)(A)—Materials of construction;
- 1910.119(d)(3)(i)(B)—Piping and instrument diagrams (P&ID's);
- 1910.119(d)(3)(i)(C)—Electrical classification;

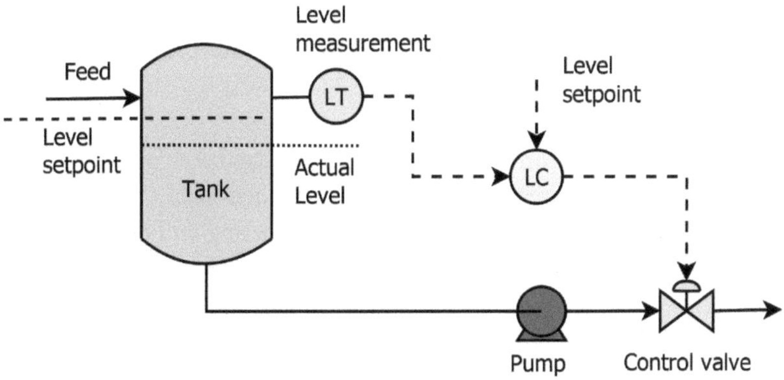

Fig. 2.8 A sample piping and instrument diagram.

- 1910.119(d)(3)(i)(D)—Relief system design and design basis; 1910.119 (d)(3)(i);
- (E)—Ventilation system design;
- 1910.119(d)(3)(i)(F)—Design codes and standards employed;
- 1910.119(d)(3)(i)(G)—Material and energy balances for processes built after May 26, 1992;
- 1910.119(d)(3)(i)(H)—Safety systems (e.g. interlocks, detection or suppression systems);
- 1910.119(d)(3)(ii)—The employer shall document that equipment complies with recognized and generally accepted good engineering practices; and
- 1910.119(d)(3)(iii)—For existing equipment designed and constructed in accordance with codes, standards, or practices that are no longer in general use, the employer shall determine and document that the equipment is designed, maintained, inspected, tested, and operating in a safe manner.

Some data on hazardous materials, regulational and geographical data are collected and provided by government institutions to ease PSM for employers and employees. An example of such information system is CAMEO (Computer-Aided Management of Emergency Operations) software provided by the U.S. Department of Commerce, National Oceanic and Atmospheric Administration (CAMEO Software Suite, 2020).

All software components work interactively to display critical information in an easy-to-understand manner. It is possible to use the suite to:

- manage data for emergency planning and response (including facilities, chemical inventories, contact information, and response resources);
- access chemical property and response information;

- find out how chemicals could react if they mixed;
- estimate threat zones for hazardous chemical releases, including toxic gas clouds, fires, and explosions; and
- map threat zones and other locations of interest.

The software allows information search in the databases on safety topics, modeling hazards, and emergency planning. Unfortunately, CAMEO does not provide complex big data analyzing and predicting and preventing hazards, although it is still a very helpful tool for emergency response.

2.2.2 HAZOP

HAZOP (hazard and operability studies) is a structured collective process of detailing, and identification of hazards and problems in system performance, completed by a group of experts. The HAZOP study is designed to identify potential deviations from the design intent, assess their possible causes and assess their consequences. A qualitative approach is similar to quantitative analysis, except that a qualitative approach requires less detail and does not require significant time. Although the results of this method are not as accurate as the results of quantitative analysis, it provides a basis for determining priorities for a risk-based inspection program (*IEC61882:2002 Hazard and operability studies (HAZOP studies)—Application Guide. British Standards Institution*, 2002; Nolan, 1994).

One of the intents of any industrial plant is to operate in the safest and the most efficient way (*ISO/TR 12489:2013. Petroleum, petrochemical and natural gas industries—Reliability modeling and calculation of safety systems*, 2013; IEC 31010:2019, Risk management—Risk assessment techniques, 2019). When industrial equipment is designed, assembled and working together, every item of a plant, each pump, valve, or pipeline, will need to function properly. For example, let us consider a cooling water system as a part of an

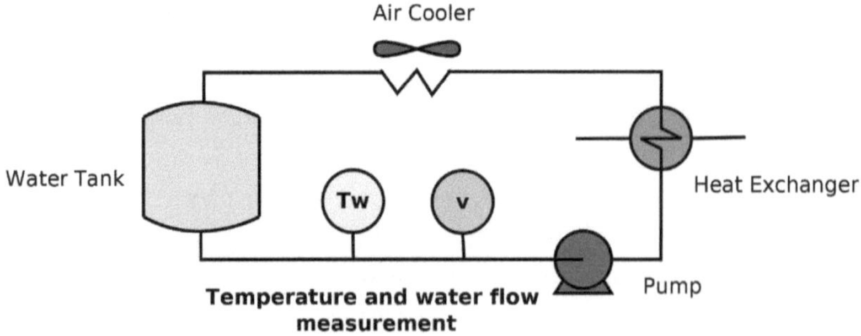

Fig. 2.9 Cooling water system.

overall design. The main parts of this facility are the cooling water circuit pipework with the installed pump, fan cooler, and heat exchanger (Fig. 2.9).

The design intent of this small part of the plant is to continuously circulate cooling water at an initial temperature of T_W (°C) and at a rate of v (liters per hour). It is usually at this low level of design intent that a HAZOP study is conducted. A deviation from the design intent in the case of the cooling facility would be a cessation of circulation or the water being at too high an initial temperature. In the case above, failure of the pump would be a cause of the deviation. The consequences could be overheating of other parts of the plant, explosion, or failure of the entire process.

Essentially the HAZOP procedure involves taking a full description of a process and systematically questioning every part of it to establish how deviations from the design intent can arise. Once identified, an assessment is made as to whether such deviations and their consequences can have a negative effect upon the safe and efficient operation of the plant. If considered necessary, action is then taken to remedy the situation. This critical analysis is applied in a structured way by the HAZOP team, and it relies upon them releasing their imagination in an effort to discover credible causes of deviations.

Table 2.1 Typical keywords and example.

Keyword	Meaning	Process industry example
No	The design intent does not occur (e.g., Flow/No), or the operational aspect is not achievable (Isolate/No)	No part of the intention is achieved (e.g., no flow, no isolation)
Less	A quantitative decrease in the design intent occurs (e.g., Pressure/Less)	A quantitative decrease (e.g., lower temperature)
More	A quantitative increase in the design intent occurs (e.g., Temperature/More)	A quantitative increase (e.g., higher pressure)
Reverse	The opposite of the design intent occurs (e.g., Flow/Reverse)	Reverse flows in pipes and reverse chemical reactions
Also	The design intent is completely fulfilled, but in addition some other related activity occurs	Flow/Also indicating contamination in a product stream, or Level/Also meaning material in a tank or vessel which should not be there

Continued

Table 2.1 Typical keywords and example—cont'd

Keyword	Meaning	Process industry example
Other	The activity occurs, but not in the way intended	A result other than original intention is achieved (e.g., Flow/Other could indicate a leak or product flowing where it should not, or Composition/Other might suggest unexpected proportions in a feedstock)
Fluctuation	The design intention is achieved only part of the time	An airlock in a pipeline might result in Flow/Fluctuation
Early	Usually used when studying sequential operations, this would indicate that a step is started at the wrong time or done out of sequence	Something happens early relative to dock time, e.g., cooling or filtration
Late	As for Early	Something happens late relative to dock time, e.g., heating

An essential feature in this process of questioning and systematic analysis is the use of keywords to focus the attention of the team upon deviations and their possible causes. Typical keywords and examples are presented in Table 2.1.

The flow chart of HAZOP study is presented in Fig. 2.10.

In simple terms, the HAZOP study process involves applying in a systematic way all relevant keyword combinations to the plant in question in an effort to uncover potential problems. The results are recorded in a specific columnar format with standardized headings in Table 2.2.

Any existing protective devices that can prevent the cause or safeguard against the adverse consequences should be recorded in the column "Safeguard." In case a credible cause leads to a negative consequence, it must be decided whether some action should be taken. For instance, it may be recommended to analyze the issue of the effectiveness of the leak detection system in the gas terminal system when using the bypass line and cutting off the terminal from the main gas pipeline.

The HAZOP report is a key document pertaining to the safety of the plant (project). As a result, it usually states the number of recommended actions. The HAZOP report is compiled as soon as possible after the end

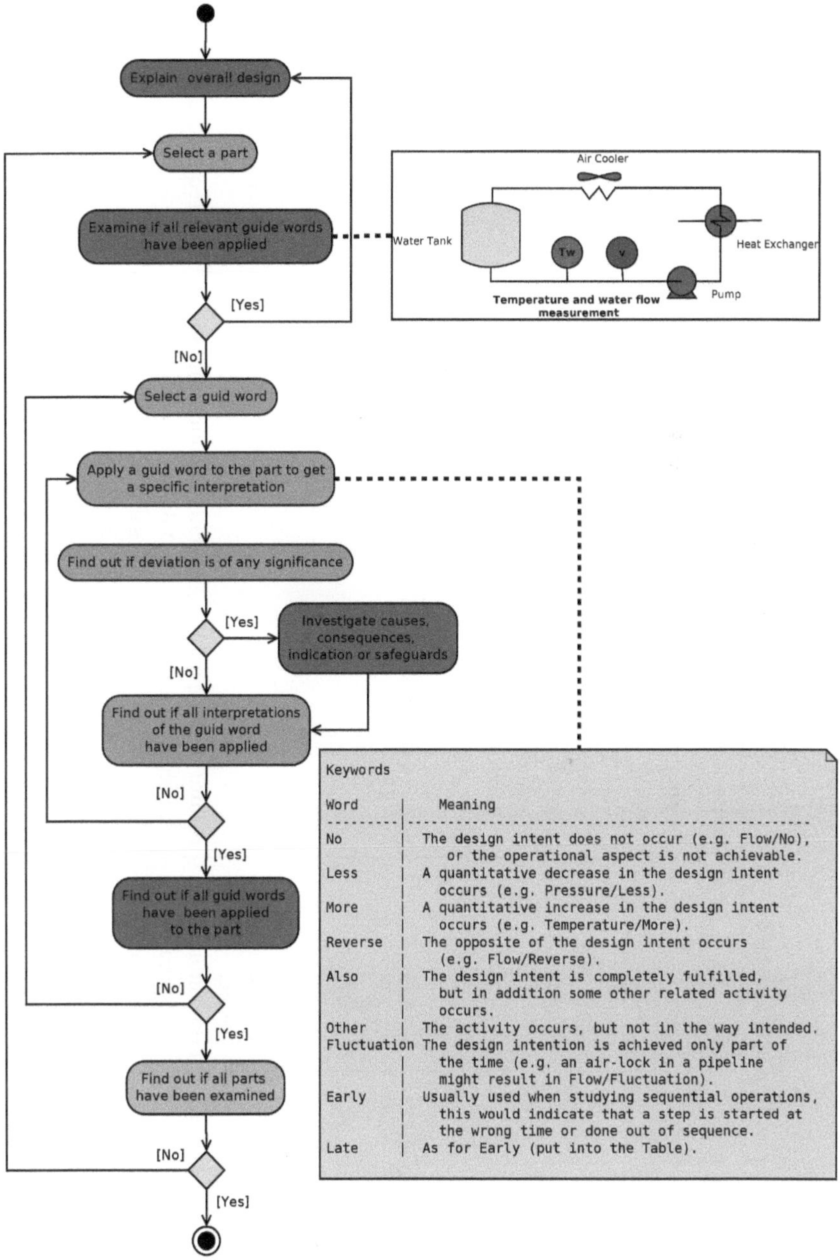

Fig. 2.10 The HAZOP process flow chart.

Table 2.2 The fragments of HAZOP work table.

Keyword	Deviation	Cause	Consequence	Safeguards	Action
No	No gas flow	Pipeline rupture. Input collector closed	Stopping gas supply to the consumer. Emergency shutdown of power supply gas generators. Economic losses	Pipeline leak detection system and actions to cut off the emergency section (−30 km). Using an emergency uninterruptible power supply (UPS). Locking taps in the open state	To analyze the issue of the effectiveness of the leak detection system in the terminal system when using the bypass line and cutting off the Terminal from the main gas pipeline
Reverse	Reverse gas flow	Opening the pressure relief line from the filter until the valves are closed at the outlet	Filter destruction	Switching to the second spare filter	Analyze design decisions on the consequences and the possibility of increased filter protection during reverse flow
Less	Temperature reduction	Low ambient temperature. Throttle effect	Failure of equipment, instrumentation and automation	Climatic conditions are taken into account by calculation. The filter housing material is designed for temperature minus 60°C	Not required

More	Level up	Level sensor failure	*Separator filter overflow and liquid getting into the gas pipeline downstream. Emergency Stop. **Drain tank overflow	*Duplication of level control by a backup sensor and visually. **Extremely high level alarm and dipping flow automatic valve at the inlet to the filter line	*Consider a diagram of the cause-effect relationships of an emergency stop with a new design. **Clarify the algorithm of emergency protection when considering design solutions for a gas filtration technological system

of the study, and once completed does not change. On the other hand, the Action File is only started at the end of the study, and its contents will continue to change perhaps for many months, until the very last action has been reviewed and accepted as having been satisfactorily discharged.

Essentially, this Action File is a binder, or perhaps some electronic form of storage. Initially, at the end of the study, it will be empty. As completed and signed, Action Response Sheets are returned, and are housed in the Action File. Periodically the returned responses will be input into the data file (either manually or electronically, according to the system being used). When all action responses have been reviewed and accepted, it finally becomes a static record containing the complete history of the implementation of the HAZOP Study's findings.

Currently, most of the expert procedures are not automated at all, or are poorly automated. In the world of big data, using intellectual expert systems for choosing keywords, main plant parts, typical deviations, causes, consequences, safeguards, and action can significantly reduce time for HAZOP procedures, reduce the number of HAZOP team members, and provide wider and deeper research for every new project designed (Crowley, 2018; Zhao, Bhushan, & Venkatasubramanian, 2005).

In detail, the modernized HAZOP technique based on information technologies and the big data paradigm allows shorter (about 40% time reduction), more focused HAZOP full team sessions, and provides improvements in knowledge and approach to HAZOP to be reflected in the HAZOP report and Action File. In all, it may reduce safety external costs by more than 25% (Crowley, 2018).

On the other hand, in spite of information technologies implementation, full team meetings will always be important in order not to miss any important issue by relying also on HAZOP experts' knowledge and experience.

Computer-assisted case generation gives the experts the ability to implement high-definition models of the plants, and their interactions based on chemistry, physics, thermodynamics, and chemical behavior. This allows prediction of unacceptable deviations and risk analysis (Fig. 2.11).

It gives the ability to obtain good predictions of chains of events, and their outcomes to get a good vision of the highest risks, the way that safeguards influence the outcome of events, and how this reduces the risk.

In addition, implementing Internet of Things (IoT) and big data technologies provides using actual plant and safeguards performance in the real-time mode to validate the models during the plant operation.

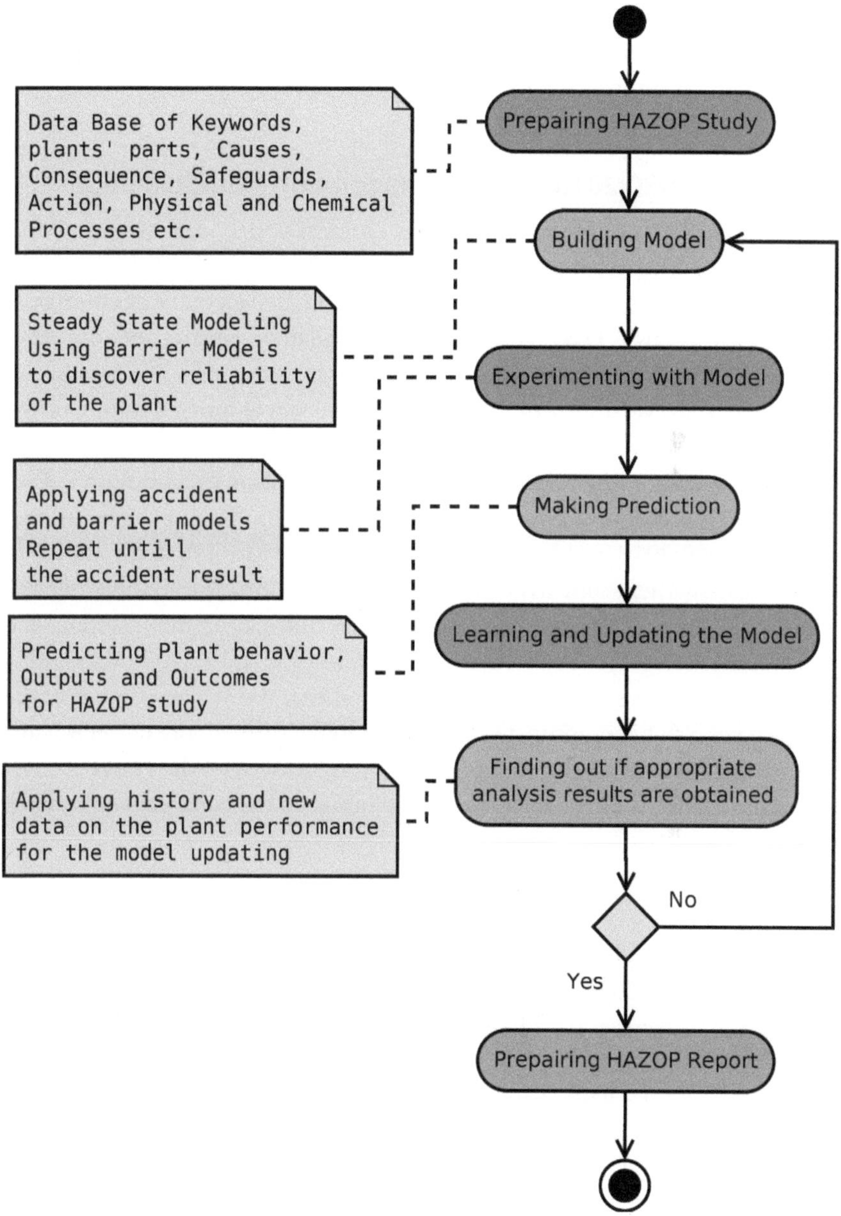

Fig. 2.11 Model-based advanced HAZOP study.

Having all these in mind, it can be concluded that traditional formal HAZOP meeting may become not a design tool only but also a confirmation tool based on information technologies' aid.

2.2.3 ISO 31000:2018—Risk management—Guidelines

This standard was prepared by the international organization ISO (International Organization for Standardization) and is one of the key documents regulating the risk-based approach in the management of enterprises and organizations. Two years ago, an updated version of the ISO 31000:2018 standard was adopted to replace the previous version of 2009. The new version took into account the development trends of the modern market, new problems that organizations and businesses face, as well as new opportunities opened up by information technology (ISO 31000:2018, Risk management—Guidelines, 2018).

ISO 31000:2018, unlike ISO 31000: 2009, provides a strategic approach to risk management and pays more attention to both the involvement of senior management and the integration of the risk management system into the overall management system of the organization (ISO 31000 standard: a different perspective on risk and risk management, n.d.). The content of the standard has also been optimized in order to reflect the model of open management systems that regularly exchange information with the external environment. Fig. 2.12 shows the relationship between the basic concepts of risk

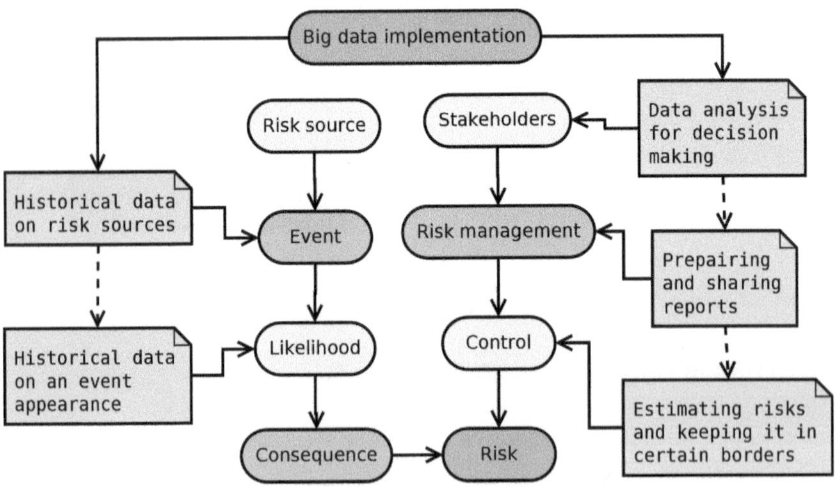

Fig. 2.12 Basic concepts of risk management and their relationship with big data.

management in terms of the application of information technologies and big data technologies.

The standard contains recommendations for organizations on the implementation of risk-based decision-making in enterprise management processes, namely planning, implementation of plans, reporting, enterprise policy, and organization culture and values. As a global goal, ISO 31000:2018 defines the development of a risk management culture in which stakeholders and employees are informed of the critical importance of risk monitoring and management. This standard applies to all types of risk faced by a company or enterprise in the face of uncertainty and which must be minimized in order to increase the efficiency of the organization. Since process safety is a fundamental component of the reliable functioning of any production enterprise, the ISO 31000:2018 standard is necessary for consideration by the enterprise management, middle managers, as well as engineers and security specialists. The standard has retained its previous components and also governs the basic principles, structure, and process of risk management.

Among the principles of risk management, we would like to dwell on principle (f): "Best Available Information." This principle means that information should be timely, clear, and accessible to all interested parties. Baseline data for risk management is based on information about the past and current information, as well as forecasts. Risk management explicitly takes into account any limitations and uncertainties associated with such information and the accuracy of forecasting.

As part of the risk management process, from the point of view of the application of big data, the subprocesses (6.2) "Communications and consultations," (6.6) "Monitoring and analysis," and (6.7) "Registration of results and reporting" should be distinguished.

The purpose of communication and consultation is to help relevant stakeholders understand the risks, the basis for decision-making, and situations requiring specific action. Communications are aimed at raising awareness and understanding of risk, while consultations involve obtaining feedback and information for informed decision-making. Close coordination between them should facilitate the timely, relevant, accurate, and understandable exchange of information, taking into account issues of confidentiality and integrity of information, as well as the right to privacy.

Communication and consultation with relevant external and internal stakeholders should be carried out at all stages of the risk management process. Communication and consultation solve the following problems:

- combining different areas of knowledge at each stage of the risk management process;
- ensuring that, when defining risk criteria and assessing risks, various points of view are taken into account accordingly;
- providing sufficient information to ensure risk control and decision-making; and
- the formation of a sense of involvement among those at risk.

The purpose of *monitoring and analysis* is to ensure and improve the quality and effectiveness of the development and implementation of the process, as well as its results. Continuous monitoring and periodic analysis of the risk management process and its results should be a planned part of the risk management process with clearly defined responsibilities. Monitoring and analysis should be carried out at all stages of the process. Monitoring and analysis includes planning, collecting, and analyzing information, recording results, and providing feedback. The results of monitoring and analysis should be taken into account in the management of the organization, during measurements and reporting.

Registration of results and reporting. The risk management process and its results should be documented and disseminated by appropriate methods. Registration and reporting have the following objectives:

- dissemination of information about the activities and results of risk management throughout the organization;
- providing information for decision-making;
- improvement of risk management activities; and
- assistance in interaction with stakeholders, including those responsible for the results and implementation of risk management actions.

Decisions regarding the creation, storage, and processing of documented information should take into account possible use (accessibility as required) and the confidentiality of information.

Reporting is an integral part of the organization's management and should improve the dialogue with stakeholders and support senior management and oversight bodies in fulfilling their responsibilities.

Factors to consider when reporting include but are not necessarily limited to:

- various stakeholders and their specific needs and requirements related to information;
- reporting costs, frequency, and timeliness;
- reporting method; and
- the importance of information in terms of organizational goals and decision-making.

Concluding a brief analysis of the ISO 31000:2018 standard, we find that the success of risk management largely depends on the amount of information available, its quality, the timeliness of receipt, and the effectiveness of big data processing algorithms.

2.2.4 ISO/IEC 31010-2019

This document provides guidance on the selection and application of various methods (techniques) that can help improve the way uncertainty is taken into account and help understand risk. In other words, if ISO 31000:2018 offers a strategic approach to risk management as a whole (ISO 31000:2018, Risk management—Guidelines, 2018), then IEC 31010:2019 equips the user with specific modeling techniques and tools to take into account uncertainties in particular risk identification, analyzing, evaluating, and decision-making for risk managing. Methods can be adapted, combined, and applied in a new way if necessary (IEC 31010:2019, Risk management—Risk assessment techniques, 2019).

The definitions of such key terms as risk, likelihood, and probability were given in Section 2.1.2—Risk Definition and Risks Calculation. In addition, let us consider some more necessary terms (IEC Electropedia, n. d.; ISO Guide 73:2009, Risk management—Vocabulary, 2009):

- *Opportunity* is a combination of circumstances that are expected to be conducive to goals.
- *Risk driver* is a factor that has a significant impact on risk.
- *Threat* is a potential source of danger, harm, or other undesirable consequences.

The core concept of IEC 31010:2019 is *uncertainty*, which was considered in detail in the first chapter of this book. Uncertainty is in direct relation to information (or lack of information) and therefore, with data and big data. We shall now analyze the selected types of uncertainty that can be reduced with big data.

In general, this is uncertainty, which, as a rule, is the result of a lack of knowledge and, therefore, can be reduced by collecting more data, refining models, improving sampling methods, etc. Commonly recognized types of uncertainty are the following:

- *Linguistic uncertainty* is caused by the ambiguity inherent in spoken languages. Here, intelligent speech recognition and translation systems can be used.
- *Uncertainty about the truth* of the assumptions, including assumptions about how people or systems can behave. In this case, more gathered statistics on people or equipment behavior is useful.

- *Variability of parameters* on which the solution should be based. Statistics theory can help to solve the problem.
- *Uncertainty about the reliability or accuracy of the models* that were created to predict the future. New types of models and predictive analytics tools based on big data technologies should be used to improve the situation.
- *Uncertainty arising from the limitations of the human mind*, for instance, in understanding complex data, predicting situations with long-term consequences, or making impartial judgments. For this purpose, intellectual control and management systems can be suggested.

Recognizing that there is a certain type of uncertainty allows early warning systems to be introduced in order to detect changes in the system in a proactive manner and to take measures to increase resilience to cope with unexpected circumstances.

As we have already mentioned, risk includes the effects of any of the types of uncertainty on objectives. The uncertainty may lead to different consequences. The consequences may not be noticeable or measurable at first, but may accumulate over time. Sources of risk may include intrinsic variability or uncertainties associated with a variety of factors, including human behavior and organizational structures, for which it may be difficult to predict any specific event that may occur. Thus, risk cannot always be easily calculated based on a series of events, their consequences, and probabilities.

Additional information and special techniques may be required to combat uncertainties. This information should provide the input data for models, statistical analysis or methods described in the standard, and decision-making.

When it comes to big data, the issues of information collecting, storing, processing, and providing access become even more relevant, and this scheme needs to be worked out before proceeding with risk assessment. On the other hand, one should consider the form in which the results of the assessment should be stored, as well as how these records should be created, stored, updated, and made available to potential users of information. It is important that the sources of information are always indicated. No information of unknown origin should remain. On the one hand, this helps to ensure information security, and on the other hand, it will allow the fullest use of information. Another important issue is the choice of sources of information to reduce the number of "white spots" in the information picture of the assessed situation. The main possible sources of information, common

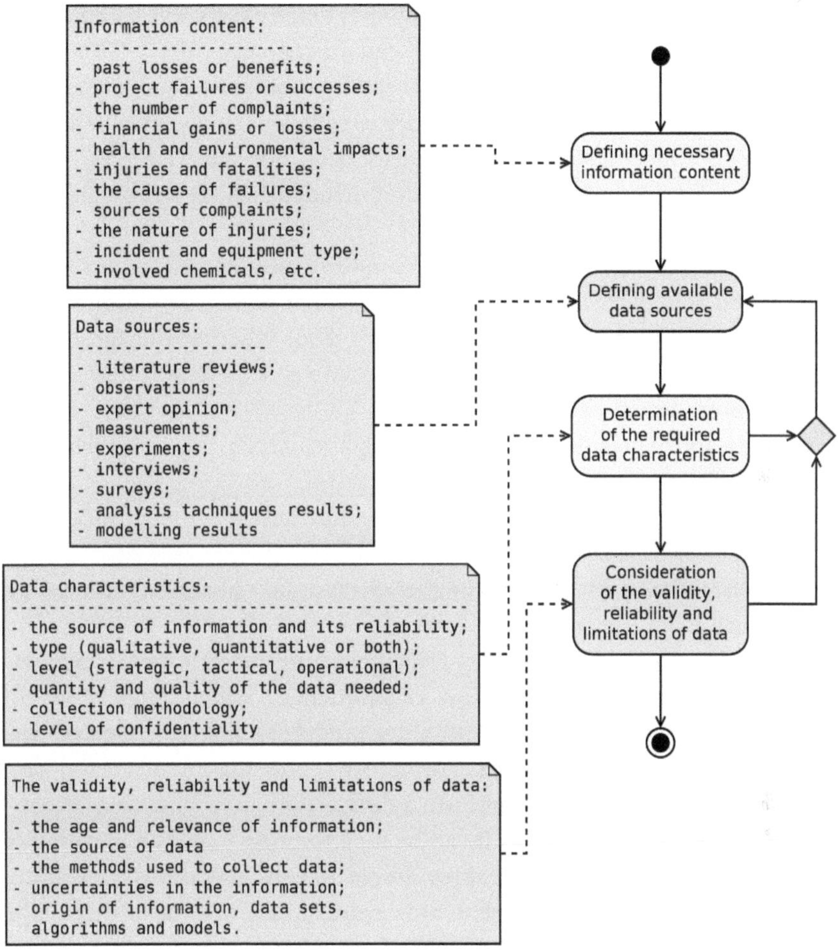

Fig. 2.13 The collecting data process for risk assessment.

information content, and data characteristics for the data collecting process for risk assessment are presented in Fig. 2.13.

We should remember that it can be an iterative process. For instance, while considering the validity, reliability, and limitations of data, we may need to come back to determination of the required data characteristics or defining available data sources in case the characteristics or sources do not meet the reliability requirement. When we deal with data obtained from measurements, experiments, or simulation results, especially in real time, a number of tasks immediately arise for collecting, transmitting, and storing information specific to big data technologies. These tasks and technologies will be discussed in more detail in subsequent chapters.

It is also necessary to keep in mind that when data for subsequent analysis are obtained from a sample, it is necessary to indicate the statistical reliability necessary to collect a sufficient amount of data. For small samples, special methods of data analysis can also be recommended, which will be discussed later in this book.

If there are data or the results of previous risk assessments, it must first be determined whether there has been any change in the conditions under which the assessment was carried out and, if so, whether the earlier data or results remain relevant. Thus, usually data are collected for future analysis that can provide:

- trends and patterns, including periodicity, which indicate what may affect future events;
- correlations that may indicate possible cause and effect relationships for further verification;
- understanding past event consequences and their likelihood in purpose to learn from experience;
- identifying and understanding limitations and uncertainties in data.

IEC 31010:2019 standard also recommends wide application of modeling techniques including computer models to analyze collected data for the risk assessment. A model is a simplified representation of a real system or process. Its purpose is to transform a difficult situation into simpler concepts that are easier to analyze. It can be used to understand the meaning of data and to simulate what can happen in practice under various conditions. Software can be used to represent and organize data or to analyze it. The software used for modeling and analysis often provides a simple user interface and quick output, but these characteristics can lead to incorrect results that are invisible to the user. Data entry errors are often detected when input changes. This approach also provides information on the sensitivity of the model to data changes. A good understanding of the mathematics related to a particular analysis is recommended to avoid erroneous conclusions. Not only are the above errors likely, but the choice of a specific program may not be appropriate. It is easy to follow the program and assume that the answer is correct. Evidence needs to be gathered to ensure that the results are reasonable. Most of the techniques suggested in the standard suppose implementation of models and various software products too.

After all, risk assessment techniques are designed to help staff and stakeholders identify and analyze uncertainties and associated risks in this broad, diverse, and complex sense in order to support more informed decisions and actions. Next, we look at how techniques can be categorized, including on the basis of the possible use of information and big data technologies.

Characteristics of techniques can be used to select methods to apply. These characteristics are application, scope, processes or equipment level, time horizon, decision level, starting info/data needs, specialist expertise, qualitative/quantitative, effort to apply.

From a big data point of view, we are mostly interested in "starting info/data needs." This states the level of starting information or data needed to implement a certain technique. The gradations are "high," "medium," and "low." In addition, we should pay more attention to techniques closer to process safety issues. Most of the techniques used for *risk analysis* require high or medium data volumes, i.e., understanding consequence and likelihood in the framework of the ISO 31000 process. These techniques include:

- Bayesian analysis;
- Bayesians networks;
- cause-consequence analysis;
- event tree analysis (ETA);
- fault tree analysis (FTA);
- Markov analysis; and
- Monte Carlo analysis.

Another group of high or medium data methods is aimed at *risk identification*. The techniques are:

- checklists classification;
- failure modes and effects (and criticality) analysis (FME(C)A);
- hazard and operability studies (HAZOP);
- scenario analysis; and
- structured "what-if" technique (SWIFT).

Bayesian analysis (Ghosh, Delampady, & Samanta, 2006) enables both data and subjective information to be used in making decisions. Observed event data can be combined with the prior distribution through a Bayesian analysis to provide a posteriori estimation of the risk parameters. Input data are the judgmental and empirical data needed to structure and quantify the probability model. Outputs are estimates, both single numbers and intervals, for the parameter of interest. The limitation is that solving complex problems can incur high computational costs.

Bayesian networks (BNs) (Jensen & Nielsen, 2007; Neil & Fenton, 2012) contain variables that represent uncertain events and can be applied to estimate likelihood, risk, or key risk drivers leading to specified consequences.

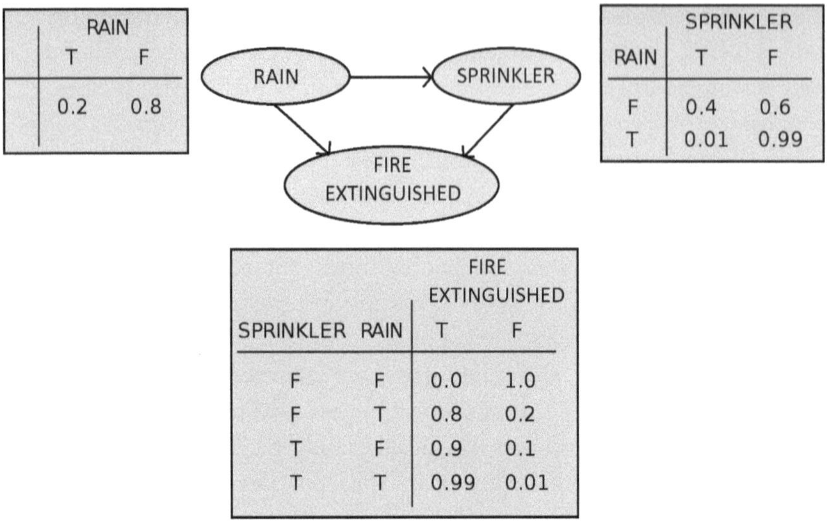

Fig. 2.14 An example of a simple Bayesian network.

BNs can be useful for mapping risk analysis for nontechnical stakeholders, contributing to the transparency of assumptions and processes and treating uncertainty in a mathematical manner. Input data for BNs are the random variables (discrete and/or continuous) (nodes), and the causal links between them and the prior and conditional probabilities for these relationships. Output data are conditional and marginal distributions in a graphical form that is generally easy to interpret (Fig. 2.14).

Limitations are as follows:

- It is difficult to determine all interactions for complex systems, and they can become computationally time-consuming when the conditional probability tables become too large.
- Setting parameters requires knowledge of many conditional probabilities, which are usually provided by an expert opinion. BNs can only provide results based on these assumptions (a limitation that is common to other modeling techniques).

Event tree analysis (ETA) (Analysis techniques for dependability—Event tree analysis (ETA), 2010) is a graphical method that represents mutually exclusive sequences of events that may occur after an initial event, depending on whether the various systems designed to change the consequences are functioning or not. A tree can be quantified to provide the probabilities of various possible outcomes (Fig. 2.15).

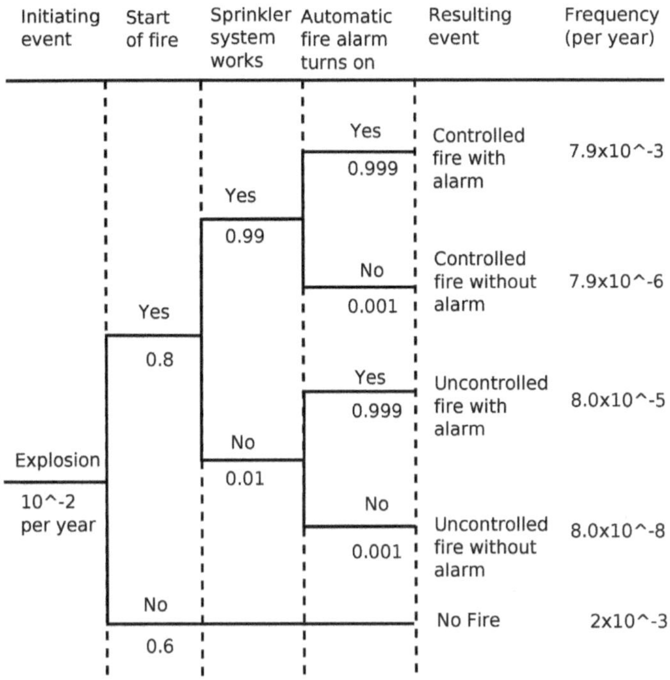

Fig. 2.15 Example of an event tree diagram for a dust explosion.

ETA can be used qualitatively to analyze possible scenarios and sequences of events after the initial event, as well as to study how various controls affect the results. The method can be used quantitatively to assess the acceptability of controls and the relative importance of different controls to the overall risk level. Inputs include a specified initiating event, information on barriers and controls, and, for quantitative analysis, their failure probabilities, as well as consideration of possible scenarios. ETA results include qualitative descriptions of potential outcomes of initiating events, quantitative estimates of the frequency or probability of events, and the relative importance of various failure sequences and associated events, as well as quantitative estimates of the effectiveness of risk control measures.

Limitations include the following. For a comprehensive analysis, it is necessary to identify all possible initiating events. There is always a chance to skip some important source events or sequences of events. Any path depends on events that occurred at previous branch points along the path. Therefore, many dependencies on possible paths are considered. However, some dependencies, such as common components, utility systems, and operators, may be overlooked, leading to optimistic estimates of the likelihood of specific consequences. For complex systems, an event tree can be difficult to build from scratch.

Fault tree analysis (FTA) (IEC 61025:2006 Fault tree analysis (FTA), 2006) is used primarily at the operational level for short and medium term issues. It is used qualitatively to identify potential causes and paths to the main event, or quantitatively to calculate the probability of a top event. Methods based on binary decision diagrams or Boolean algebra are then used to account for repeated failure modes. Input data for fault tree analysis are as follows. It is necessary to understand the system structure and functioning and the causes of failure or success in different circumstances. Detailed diagrams and manuals are useful for the analysis. For quantitative analysis of a fault tree, data are required for all base events on failure rates, or the probability of being in a failed state, or the frequency of failures, and, where relevant, repair/recovery rates. For complex situations, software usage is recommended, and knowledge of probability theory and Boolean algebra are suggested so inputs to the software are made correctly. The results of the failure tree analysis are a visual representation of how the main event can occur, which shows interacting paths, each of which includes the occurrence of two or more (basic) events; a list of minimum sets of cuts (individual paths to failure) with the data provided, and the likelihood that each of them will happen; in the case of quantitative analysis, the probability of the main event and the relative importance of the basic events are shown (Fig. 2.16).

FTA limitations include the following. In some situations, it can be difficult to determine if all the important paths to the main event are included. In these cases, it is impossible to calculate the probability of the main event. The FTA deals only with binary states (success/failure). The FTA analyzes one major event. It does not analyze secondary or random failures. An FTA can become very large for large-scale systems.

Cause-consequence analysis (CCA) (Andrews & Ridley, 2002) is used to represent the fault logic leading to a critical event, but it adds to the fault tree functionality, allowing the analysis of consecutive time outages. This method also makes it possible to take into account time delays in the analysis of consequences, which is impossible for event trees. Input data are knowledge of the system and its failure modes, and failure scenarios. The output is a schematic representation of how the system can fail, showing both the causes and consequences, as well as an estimate of the probability of occurrence of each potential consequence based on an analysis of the probabilities of certain conditions after a critical event. Limitations include the fact that CCA is

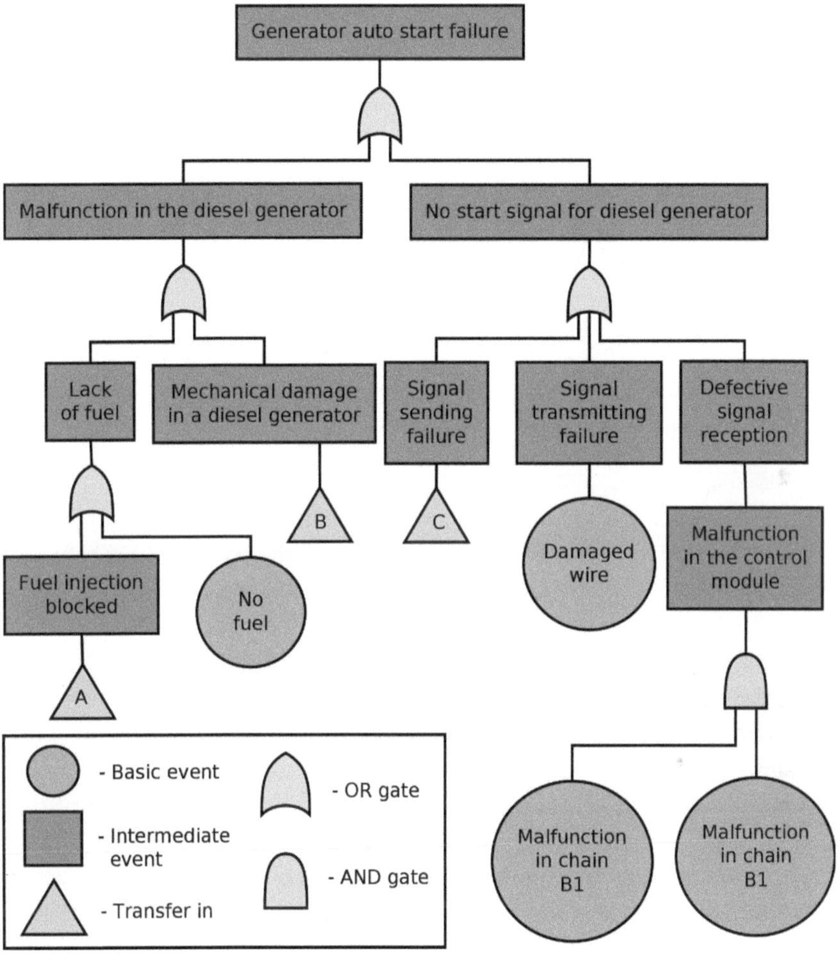

Fig. 2.16 Fault-tree diagram example.

more complex than analyzing a fault tree and an event tree, both for constructing and for the way that dependencies are processed during quantification.

Markov analysis (IEC 61165:2006 Application of Markov techniques, 2006) is a quantitative method that can be applied to any system that can be described in terms of a set of discrete states and transitions between them, provided that evolution from its current state does not depend on its state at any time in the past. It is usually assumed that transitions between states occur at specified intervals with corresponding transition probabilities (discrete time Markov chain). The inputs to a Markov analysis are a set of

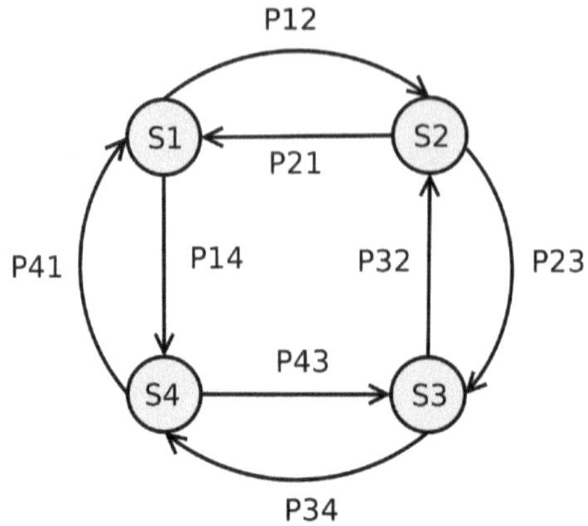

Fig. 2.17 An example of a Markov chain model.

discrete states that the system can occupy, consolidation of the possible transitions that need to be modeled, and estimates of the transition probabilities (Fig. 2.17).

As the output, Markov analysis generates estimates of the probability of the system being in any specified state. It supports many types of decisions about the types of actions that a manager can perform in a complex system (for example, to change system states and transitions between them). Limitations of the method include the following. Assumptions may not apply to all kinds of systems, in particular, the transition probabilities or transition rates between states may change over time as the system wears or adapts. Accurate modeling may require extensive data collection and verification. Too much data reduces the response to the average value.

Monte Carlo simulation (ISO/IEC & GUIDE, 2008) allows you to carry out calculations and get results when analyzing the risks associated with distributions. Modeling usually involves sampling the values of a random sample from each of the input distributions, performing calculations to obtain the value of the result, and then repeating the process through a series of iterations to build the distribution of the results. The result can be represented as a probability distribution of the value or some statistics, such as the mean value. The input data for Monte Carlo simulation are: a system model that contains the relationship between various inputs, as well as between inputs and outputs; information about the types of input or sources of uncertainty that should be presented; and the form of the required output. The output can be a single value or it can be expressed as a probability or frequency distribution, or it can be the identification of the main functions in the model that have the greatest impact on the output. The limitations associated with the data are as follows. The accuracy of the results depends on the number of simulations that can be performed. The use of the method is based on the ability to represent uncertainties in the parameters using a valid distribution. It can be difficult to create a model that adequately represents a difficult situation.

Checklists, classifications, and taxonomies can be developed for use at the strategic or operational level. They can be applied using questionnaires, interviews, structured seminars, or combinations of all three, in-person or computer-based. At the operational level, hazard checklists are used to identify hazards within the HAZID and preliminary hazard analysis (PHA) (Bailey, 1994; Popov, Lyon, & Hollcroft, 2016). These are preliminary safety risk assessments that are usually performed early in the design process. Input data are data or models from which valid checklists, taxonomies, or classifications can be compiled. The results are checklists, categories, and classification schemes, as well as consideration of risks associated with the use of lists, including (in some cases) directly lists of risks and groupings of risks. The main limitations associated with information include the following: complexity can prevent identification of relationships (e.g., relationships and alternative groupings); lack of information can lead to duplication and/or gaps (for example, schemes are not mutually exclusive and collectively exhaustive).

In the frame of *Failure modes and effects (and criticality) analysis* (FME(C)A), we subdivide a system, a process, or a procedure into elements (Failure modes and effects analysis (FMEA and FMECA, 2018). For each element, the ways in which it might fail, and the failure causes and effects are considered. The procedure can be followed by a criticality analysis, which defines the significance of each failure mode. This can be done at any level of system breakdown, from flowcharts to detailed system components or process steps. The input data includes information about the system to be analyzed and its elements in sufficient detail for a meaningful analysis of the ways in which each element can fail, and the consequences if this happens. The necessary information may include drawings and block diagrams, detailed information about the environment in which the system operates, and historical information about failures, if any. The output from the FMEA is a worksheet with failure modes, effects, causes, and existing controls, as well as the criticality criterion for each failure mode (if FMECA) and the methodology used to determine it. The limitation is that FMEA can only be used to identify individual failure modes, not combinations of failure modes, and this can be difficult for complex hierarchical systems.

Hazard and operability study (HAZOP) (*IEC 61882:2016 RLV Hazard and operability studies (HAZOP studies)—Application guide*, 2016) is a structured and systematic study of a planned or existing process, procedure, or system, which includes identifying potential deviations from the design of the project and studying their possible causes and consequences. It was analyzed in detail above in this chapter. The input includes current information about the system to be verified, as well as about the intentions and technical characteristics of the project. For industrial equipment, this may include drawings, specifications, process diagrams, control and logic diagrams, as well as operating and maintenance procedures. The results include a HAZOP meeting record with a deviation record for each viewpoint. Entries should include the keyword used and possible causes of deviations. They may also include actions to address identified problems and the person responsible for the action. Some limitations of the technique are as follows. Detailed analysis can be time consuming and, therefore, expensive. Detailed analysis requires a high level of documentation or specification of a system/process and procedure. This can focus on finding detailed solutions, rather than challenge fundamental assumptions (however, this can be mitigated by a phased approach).

Scenario analysis (Pareek, n.d.; Heijden, 1996) is a class of methods that develop models for a possible future. In general terms, it consists of identifying a likely scenario and working out what might happen in light of various possible future events. Analyzing the scenario requires data on current trends and changes, as well as ideas for future changes. The conclusion can be a "story" for each scenario, which tells how you can move from the present to the objective scenario. The effects in question can be both beneficial and harmful. Stories may include plausible details that add value to the scripts. Limitations include the following: the scenarios used may not have an adequate basis, for example, the data may be speculative, which can lead to unrealistic results that cannot be recognized as such; there is little evidence that the scenarios studied in the long run are those that actually occur.

Structured "what-if" technique (SWIFT) (Card, Ward, & Clarkson, 2012) uses a structured brainstorming session in an organized workshop where a predefined set of guiding words (time, quantity, etc.) are combined with clues received from participants, which often begin with phrases such as "what if?" or "how could?" SWIFT is a high-level risk identification method that can be used independently or as part of a phased approach to increase the effectiveness of bottom-up methods such as HAZOP or FMEA.

The technique can be applied to systems, plant elements, procedures, and plants in general. The input provides a clear overview of the system, procedure, installation object and/or change, and external and internal contexts. The output includes a risk register with activities or tasks ranked by risk that can be used as the basis for a treatment plan. Limitations include the following: if the workshop team does not have a sufficiently broad base of experience or the hint system is not comprehensive, some risks or dangers may not be identified; applying a high-level technique may not reveal complex, detailed, or interrelated causes.

2.3 Standards and big data

As already shown in the previous paragraphs, modern standards in the field of safety, in particular, process safety, are becoming more focused on the widespread use of information, information technology, including big data technologies. Analyzing the content of the considered standards, we can ultimately identify several main classes of information used (Fig. 2.18).

As can be seen from the figure, different standards and regulations require the use of similar information, for example, information about the

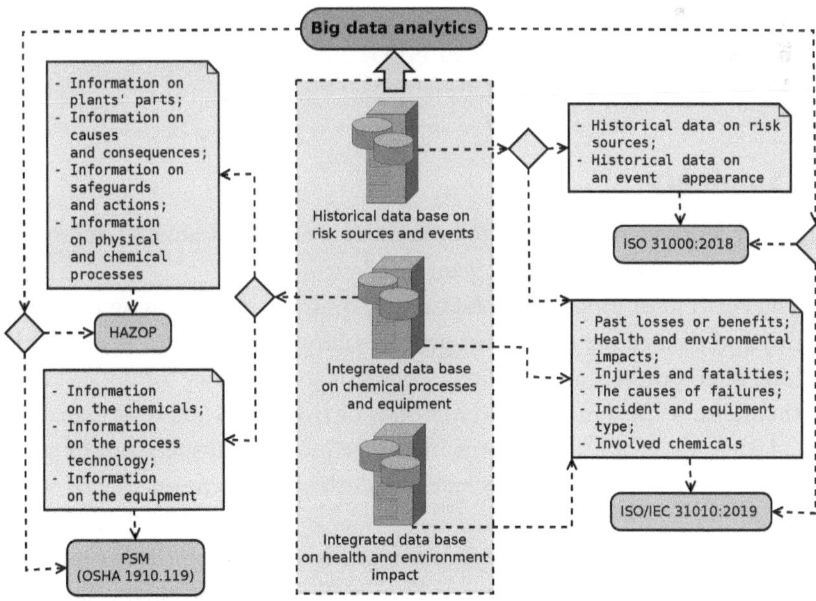

Fig. 2.18 Main classes of big data used in process safety standards.

equipment used is required for both HAZOP analysis (*Practical Hazops, Trips and Alarms*, 2004), the implementation of the PSM procedure.

(Sutton, 2010), and for risk assessment according to ISO/IEC 31010: 2019 (IEC 31010:2019, Risk management—Risk assessment techniques, 2019), which uses information on the equipment used and chemical processes. Each of the enterprises can choose the most suitable safety management scheme in accordance with the goals and available organizational, technical, and financial resources. At the same time, from an information point of view, no matter what standards an enterprise follows, there can be one source of data on equipment and technological processes. Such a source can be a single database containing information from various enterprises and industry organizations. This can be both archived data and information obtained in real time.

On the other hand, in the future, big data can be successfully used to monitor compliance with standards. With a broader review, one can imagine the possibility of creating a unified global information system of standards in the field of security, containing information on all aspects of process safety management, providing a unified form of input-output of documents regulated by standards, as well as recommendations on compliance with standards based on data mining, stored in the system. Of course, this task is very complex and requires the attraction of a huge amount of material, organizational, and computational resources. The process model of such an information system, taking into account the recommended NIST big data reference architecture, (Big Data Taxonomies, 2015) is presented in Fig. 2.19.

2.4 Summary

Currently, big data is beginning to be actively used for risk analysis of process safety based on a number of relevant standards.

The concept of probability plays an important role in assessing potential hazards and risks in the context of process safety management. We considered the main points of probability theory that are necessary for understanding the mathematical models and methods of risk theory. Statistical analysis should identify significant patterns in the frequency and spectrum of accidents and use these patterns in practice to calculate the frequency of accidents of the investigated object.

One of the main areas of application of big data in the field of process safety is the assessment of process risks in order to reduce them. The risk assessment includes identifying risks, analyzing risks, and evaluating risk.

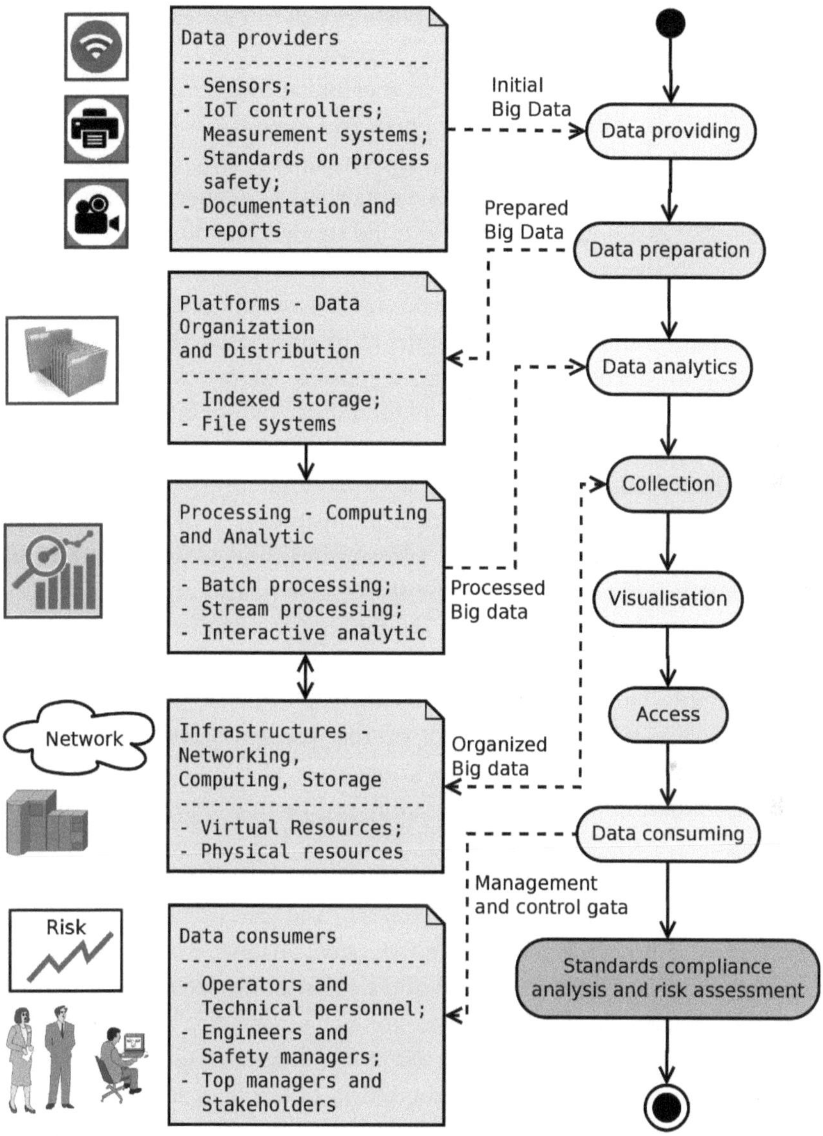

Fig. 2.19 The process model of the information support system for standardization in the field of process safety based on big data.

Risk analysis is the procedure that helps in understanding the nature of the risks and their features, including, when necessary, the level of risk. It also allows us to consider the specifics of the development of various types of accidents, as well as the determination of quantitative indicators of the associated social, material, and environmental damage.

A variety of hazard manifestations corresponds to a variety of assessments. Depending on the operating mode of the research object and risk assessment associated with the consequences of an accident at the facility. The latter are called emergency risk assessments. These two types of risk are sometimes called real and potential. Thus, risk assessment is a very complex process from an information point of view. To obtain risk assessments, it is necessary to collect a large amount of data and build models based on these data.

When calculating and predicting the process safety risks, big data can be used for dynamic risk assessment and automatic report generation. This enables hazardous production personnel to identify safety threats at an early stage, taking the necessary preventive measures to address them. In this way, a number of incidents involving the sudden shutdown of industrial plants or accidents can be avoided.

Compliance with standards plays a key role in ensuring process safety. The Occupational Safety and Health Administration (OSHA) was created by the U.S. government to ensure safe and healthful working conditions for working people by setting and enforcing standards and by providing training, outreach, education, and assistance. One of the key documents for chemical process safety is OSHA 1910.119 "Process safety management of highly hazardous chemicals." Process safety information can be considered a very important part of the process safety management (PSM). It provides personnel with knowledge of how to act to prevent hazardous situation taking into account production equipment and the processes.

The HAZOP study is designed to identify potential *deviations* from the design intent, assess their possible *causes*, and assess their *consequences*. A qualitative approach is similar to quantitative analysis, except that a qualitative approach requires less detail and does not require significant time. Although the results of this method are not as accurate as the results of quantitative analysis, it provides a basis for determining priorities for a risk-based inspection program. In the world of big data, using intellectual expert systems for choosing keywords, main plant parts, typical deviations, causes, consequences, safeguards, and action can significantly reduce time for HAZOP procedures, reduce the number of HAZOP team members, and provide wider and deeper research for every single new project designed.

An updated version of the ISO 31000:2018 standard was adopted to replace the previous version of 2009. The new version took into account the development trends of the modern market, new problems that organizations and businesses face, as well as new opportunities opened up by information technology. The content of the standard has also been optimized in order to reflect the model of open management systems that regularly exchange information with the external environment. As part of the risk management process, from the point of view of the application of big data, the subprocesses "Communications and consultations," "Monitoring and analysis," and "Registration of results and reporting" should be distinguished.

If ISO 31000:2018 offers a strategic approach to risk management as a whole, then IEC 31010:2019 equips the user with specific modeling techniques and tools to take into account uncertainties in particular risk identifying, analyzing, evaluating, and decision-making for risk managing. Sources of risk may include intrinsic variability or uncertainties associated with a variety of factors, including human behavior and organizational structures, for which it may be difficult to predict any specific event that may occur. Thus, risk cannot always be easily calculated based on a series of events, their consequences, and their probabilities. It may need some additional information and special techniques to fight uncertainties. This information should provide the input data for models, statistical analysis, or methods described in the standard, and decision-making.

The modern standards in the field of safety, in particular, process safety, are becoming more focused on the widespread use of information, information technology, including big data technologies. At the same time, from an information point of view, no matter what standards an enterprise follows, there can be one source of data on equipment and technological processes. Such a source can be a single database containing information from various enterprises and industry organizations. This can be both archived data and information obtained in real time.

The following chapters will examine in detail the technologies needed to apply big data in standards-based risk management.

2.5 Definitions

Risk is an effect of uncertainty on objectives.

Uncertainty refers to epistemic situations involving imperfect or unknown information.

The *source of risk* is an element that can lead to a risk.

An *event* is the occurrence or change of a specific set of circumstances.

Likelihood is the chance that something happens; it is defined, measured, or determined objectively or subjectively, qualitatively or quantitatively, and described using general terms or mathematically.

Probability is a measure of the chance of occurrence, expressed as a number from 0 to 1, where 0 is impossibility, and 1 is absolute certainty.

A *control* is a measure that supports and (or) modifies risk.

Risk assessment is a term used to describe the overall process or method where hazards and risk factors are identified that have the potential to cause harm (hazard identification).

Dynamic risk analysis is an advanced methodology for risk prevention and detection that implements a distinctive approach of identifying hidden near-misses in the process data.

Process safety management system is a regulation promulgated by the U.S. Occupational Safety and Health Administration (OSHA).

Hazard and operability studies (HAZOP) is a structured collective process of detailing and identification of hazards and problems in system performance, completed by a group of experts.

Opportunity is a combination of circumstances that are expected to be conducive to goals.

A *risk driver* is a factor that has a significant impact on risk.

A *threat* is a potential source of danger, harm, or other undesirable consequences.

References

Analysis techniques for dependability—Event tree analysis (ETA). (2010). *IEC, 62502*.

Andrews, J. D., & Ridley, L. M. (2002). Application of the cause-consequence diagram method to static systems. *Reliability Engineering and System Safety*, *75*(1), 47–58. https://doi.org/10.1016/S0951-8320(01)00113-2.

Ash, R., & Doleans-Dade, C. (1999). *Probability and measure theory* (2nd ed.). Academic Press.

Bailey, K. (1994). *Typologies and taxonomies: An introduction to classification techniques*. SAGE Publications, Inc.

Big Data Taxonomies (2015). Retrieved February 18, 2020, from: http://nvlpubs.nist.gov/nistpubs/SpecialPublications/NIST.SP.1500-2.pdf.

Card, A. J., Ward, J., & Clarkson, P. J. (2012). Beyond FMEA: the structured what-if technique (SWIFT). *Journal of Healthcare Risk Management: The Journal of the American Society for Healthcare Risk Management*, *31*(4), 23–29. https://doi.org/10.1002/jhrm.20101.

CAMEO Software Suite. (2020). https://response.restoration.noaa.gov/cameosuite (Accessed 20 April 2020).

Chung, K. (2014). *A course in probability theory*. Academic Press.

Crowley, C. (2018). Optimising HAZOP in design and operations. Retrieved February 18, 2020, from: https://irfoffshoresafety.com/wp-content/uploads/2018/10/PS7-Optimising-HAZOP-Spend-in-Operations-and-Design-Conor-Crowley-Atkins.pdf.

Failure modes and effects analysis (FMEA and FMECA). (2018). *IEC, 60812*.

Fine, T. L. (Ed.). (1973). *Theories of probability* (pp. 85–117). ISBN:9780122564505. https://doi.org/10.1016/B978-0-12-256450-5.50007-8.

Guidelines for chemical process quantitative risk analysis. (1989). New York: Center for Chemical Process Safety of the American Institute of Chemical Engineers.

Guidelines for hazard evaluation procedures. 2nd Edition with worked examples. (1992). New York: Center for Chemical Process Safety of the American Institute of Chemical Engineers.

Guidelines for process safety documentation. (1995). American Institute of Chemical Engineers. Center for Chemical Process Safety.

Guidelines for safe storage and handling of high toxic hazard materials. (1988). New York: Center for Chemical Process Safety of the American Institute of Chemical Engineers.

Guidelines for technical process safety. (1987). New York: Center for Chemical Process Safety of the American Institute of Chemical Engineers.

Ghosh, J., Delampady, M., & Samanta, T. (2006). *An introduction to Bayesian analysis.* Springer-Verlag.

Heijden, K. (1996). *Scenarios: The art of strategic conversation.* John Wiley & Sons Ltd.

IEC 31010:2019. (2019). *Risk management—Risk assessment techniques.*

IEC 61025:2006. (2006). *Fault tree analysis (FTA).*

IEC 61165:2006. (2006). *Application of Markov techniques.*

IEC 61882:2016. (2016). *RLV Hazard and operability studies (HAZOP studies)—Application guide.*

IEC Electropedia (n.d.). Retrieved February 20, 2020, from http://www.electropedia.org/.

IEC61882:2002. (2002). *Hazard and operability studies (HAZOP studies)—Application guide.* British Standards Institution.

ISO 31000 standard: A different perspective on risk and risk management (n.d.). Retrieved February 20, 2020 (Original work published February 14, 2016) from http://radar-risk.com/news/ISO31000standard.

ISO 31000:2018. (2018). *Risk management—Guidelines.*

ISO Guide 73:2009. (2009). *Risk management—Vocabulary.*

ISO/IEC, & GUIDE. (2008). *Uncertainty of measurement—Part 3: Guide to the expression of uncertainty in measurement (GUM:1995)—Supplement 1: Propagation of distributions using a Monte Carlo method. Vol. 1.*

ISO/TR 12489:2013. (2013). *Petroleum, petrochemical and natural gas industries—Reliability modelling and calculation of safety systems.*

Jensen, F. V., & Nielsen, T. D. (2007). *Bayesian networks and decision graphs.* New York: Springer-Verlag.

Neil, M., & Fenton, N. (2012). *Risk assessment and decision analysis with Bayesian networks.* CRC Press.

Niccum, P. K. (2011). Twenty questions: Identify probable cause of high FCC catalyst loss. In *NPRA Annual Meeting Technical Papers. United States.*

Nolan, D. P. (1994). *Application of HAZOP and what-if safety reviews to the petroleum, petrochemical and chemical industries.* William Andrew Publishing/Noyes.

Occupational Safety and Health Administration (n.d.). Retrieved April 15, 2020, from https://www.osha.gov.

Pareek, M. (n.d.). *Using scenario analysis for managing technology risk.* Retrieved February 11, 2020 (Original work published November 1, 2012) from: https://www.isaca.org/resources/isaca-journal/past-issues/2012/using-scenario-analysis-for-managing-technology-risk.

Pariyani, A., Oktem, U. G., & Grubbe, D. L. (n.d.). *Process risk assessment uses big data.* Retrieved April 10, 2020 (Original work published 2013) from https://www.controleng.com/articles/process-risk-assessment-uses-big-data/.

Practical hazops, trips and alarms (pp. 31–64). (2004).

Process safety management of highly hazardous chemicals (n.d.). Retrieved April 21, 2020, from: https://www.osha.gov/laws-regs/regulations/standardnumber/1910/1910.119.

Protecting refractory linings in FCC units (n.d.). Retrieved April 15, 2020, from: https://catcracking.com/protecting-refractory-linings-in-fcc-units/.

Popov, G., Lyon, B., & Hollcroft, B. (2016). *Risk assessment: A practical guide to assessing operational risks*. Wiley.

Pugachev, V. S. (1984). In V. S. Pugachev (Ed.), *Probability theory and mathematical statistics for engineers*. Pergamon.

Resolve your process risks at their initiation stage (n.d.). Retrieved April 15, 2020, from http://nearmissmgmt.com/.

Rumshiskii, L. Z. (1965). Events and probabilities. In L. Z. Rumshiskii (Ed.), *Elements of probability theory* (pp. 1–18). Pergamon. https://doi.org/10.1016/B978-1-4831-6774-9.50006-1 (chapter I).

Shenae, L., Gabriele, L., Genserik, R., & Nicola, P. (2019). Validation of dynamic risk analysis supporting integrated operations across systems. *Sustainability*, 6745. https://doi.org/10.3390/su11236745.

Sower, V. E., Jaideep, M., & Savoie, M. J. (1997). Classics in production and operations management. *International Journal of Operations & Production Management*, 15–28. https://doi.org/10.1108/01443579710157961.

Sutton, I. (2010). Technical information and industry standards. In *Process risk and reliability management* (pp. 277–300). Elsevier BV. https://doi.org/10.1016/b978-1-4377-7805-2.10005-5 (chapter 5).

The INDEPTH Data Repository (n.d.). Retrieved March 13, 2020, from: http://www.indepth-network.org/data-stats.

The National Weather Service (NWS) (n.d.). Retrieved March 15, 2020, from: https://www.weather.gov/about/.

Tools for making acute risk decisions. (1992). AlChE/CCPS.

Topographic Database (n.d.). Retrieved March 10, 2020, from: https://www.maanmittauslaitos.fi/en/maps-and-spatial-data/expert-users/product-descriptions/topographic-database.

Xuemei, Y., Shaobo, L., & Ansi, Z. (2018). Equipment condition monitoring and diagnosis system based on evidence weight. *International Journal of Online Engineering (IJOE), 143*. https://doi.org/10.3991/ijoe.v14i02.7731.

Zhao, C., Bhushan, M., & Venkatasubramanian, V. (2005). Phasuite: An automated HAZOP analysis tool for chemical processes: Part I: Knowledge engineering framework. *Process Safety and Environmental Protection, 83*(6B), 509–532. https://doi.org/10.1205/psep.04055.

CHAPTER 3

Measurements, sensors, and large-scale infrastructures

3.1 Process state data sources

The task of ensuring the safety of technological processes, taking into account the complexity of large infrastructure facilities, can be considered a task of multilevel risk management. Earlier, we examined the state space in which we can describe the dynamics of technological processes in the form of a point motion determined by a set of set values of technological process parameters $P(X)$, where $X=(x_1, x_2, \ldots, x_n)$ is the vector of parameters characterizing the state of the technological process. In the space of technological process parameters, the state of the technological process at a given time t is determined by the point $p_k(t)$ with coordinates from X. Our control object can operate in three modes:

- the normal mode—normal operation in the zone of permissible values;
- the failure mode—operation in the event of equipment failure or violation of the subsystem operation rules with the output of the characteristics of technological processes into the failure zone; and
- the emergency mode—an emergency state and exit to the hazardous area, which can lead to a critical situation.

To determine the current state of processes, we need reliable information, which can be obtained using various sensors, such as analog sensors, video cameras, or computer programs (see Fig. 3.1).

The parameters characterizing the state of physical and chemical processes include temperature, pressure, density of substances, concentration of substances, rate of entry into chemical reactors, etc. The values of these parameters can be measured by various sensors: temperature, pressure, substance concentration, fluid flow rate, etc. Some values are difficult to measure, or this process requires a lot of effort; in this case, computer-based numerical calculations can be used (Palanivelu & Srinivasan, 2020; Wang, Zheng, & Wang, 2019).

Process Safety and Big Data
https://doi.org/10.1016/B978-0-12-822066-5.00004-2

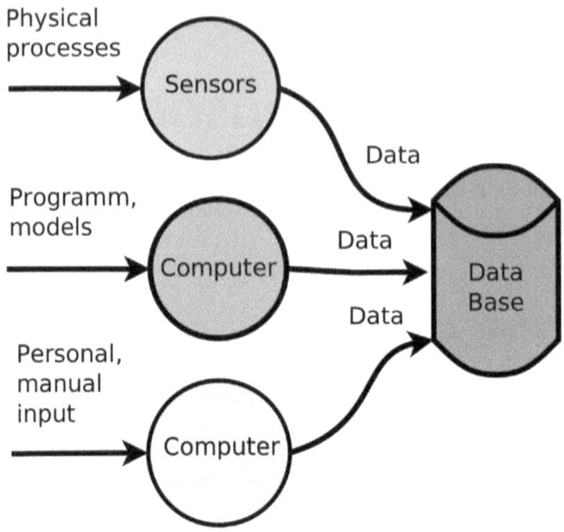

Fig. 3.1 Data sources.

The main data sources are:

- sensors that measure the parameters and characteristics of a physical object; and
- virtual sensors that calculate the parameters of a real object (for example, the level of risk). These calculations can be performed based on the digital twin (digital model).

It should be noted that when determining the values of the parameters of technological processes, errors occur during measurement, data transfer, and calculations. Metrological standards are used to unify measurement processes and analyze results (Gao, Abdul Raman, Hizaddin, & Bello, 2020; Yousefi & Rodriguez Hernandez, 2020).

Various methods are used to compensate for measurement errors, including measurement using a set of different sensors and calculation of missing data. The numerical data of the process parameters can be entered by personnel using various data input devices: keyboard, tablet, etc. In this case, data entry errors may occur due to personnel errors and limitations of the data input device.

For example, when preparing gas at gas condensate fields supplied to gas pipelines, it is necessary that the gas meets the requirements of current standards:

- Gas must not cause corrosion of pipelines, fittings, devices, etc.
- The quality of the gas must be ensured by its transportation in a single-phase gaseous state; that is, hydrocarbon liquid, water condensate, and gas hydrates should not form in the gas pipeline.

Methanol was widely used as an inhibitor for gas dehydration. Inhibition of gas hydrates is the introduction of an inhibitor into the gas stream.

Consequently, hydrocarbon and other gases are dried, and the dew point is a measure of the depth of such drying—the temperature at which drip moisture begins to form in the gas. Depending on the ambient temperature and pressure in the pipeline, the dew point changes. To determine its value, special sensors are used.

3.2 Sensors and measurements

Sensors are usually understood as devices that allow the measurement of physical quantities characterizing the state of the investigated phenomenon or process. In our case, it is convenient to understand by the term "sensor" any entity that allows us to determine the state or value of a given characteristic in any space P under study, the state of which is determined by a given set of parameters that can be determined and their values can be measured or calculated (Chohan & Upadhyaya, 1989; Zhou & Peng, 2020).

The process of obtaining numerical values characterizing the state of the technological process is presented in Fig. 3.2. The range of admissible values of P for the presented case is determined by the boundary represented by the *dashed line*.

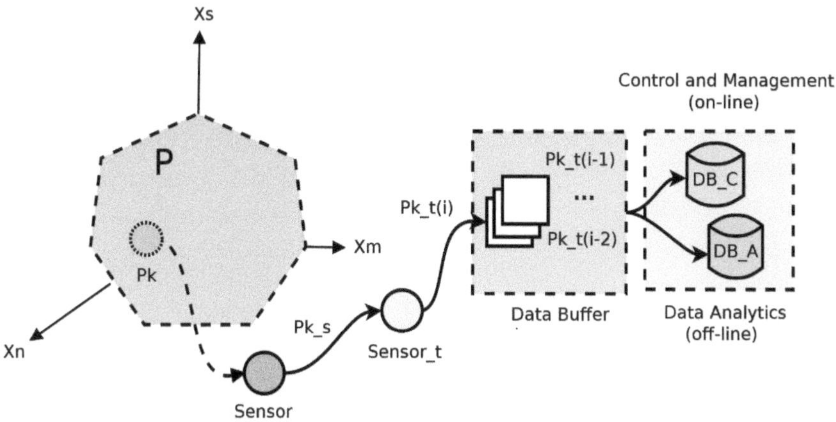

Fig. 3.2 The process of measuring the parameter of a phenomenon or process.

Using a set of sensors, the coordinates of the point $P_k = (X_{n\_k}, X_{s\_k}, X_{m\_k})$ are determined in the space of technological parameters, for example, pressure, temperature, concentration. The measured values of the physical quantities are converted into electrical signals P_{k_s}, which are then converted to digital signals $P_{k\_t}(i)$ and transmitted through the communication channel to the information processing device. In an information processing device, data is stored in a memory buffer or microcontroller memory registers. This data can be used by the controller to filter the data further, correcting possible errors.

Data from the memory buffer is transferred to the databases of a SCADA (supervisory control and data acquisition) system, and also transferred to long-term data storage for solving analysis problems. The data buffer is constantly updated, and the data is not stored in it for long. In the database of SCADA systems, data is stored for solving current management tasks and may be deleted depending on the limitations associated with the size of the database. In long-term storage databases, data can be stored for quite some time. This time is determined by the amount of incoming data and instructions for maintaining the data warehouse (Asadzadeh et al., 2020; Keren, West, Rogers, Gupta, & Mannan, 2003).

When moving from a measured parameter to its digital representation, the problem arises of representing a continuous signal on a given discrete scale. When solving it, a sampling error arises. Fig. 3.3 shows the point p in the space P; the error in determining its coordinate is determined by the value of the radius R—the smaller the value of R, the more accurate the measurement.

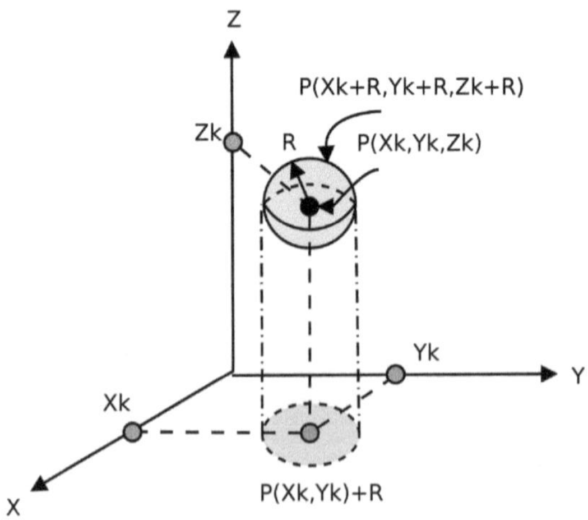

Fig. 3.3 Infrastructure space and point *p*.

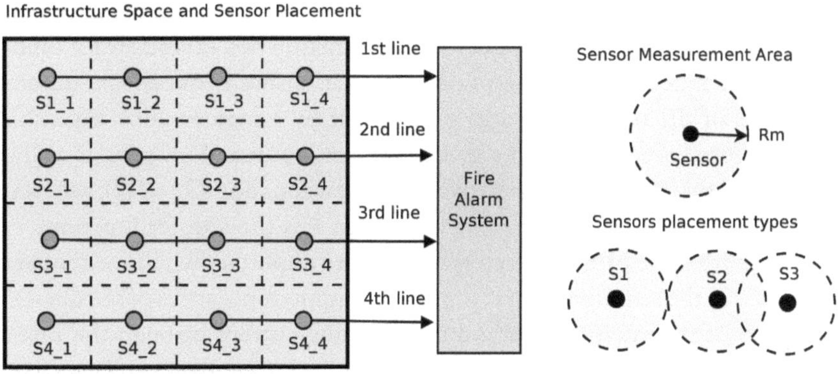

Fig. 3.4 Sensor placement and sensing area.

The placement of sensors within large infrastructure facilities is an independent design task, and in some cases, the placement rules can be set by standards. For example, the NFPA 72 standard (National Fire Alarm Code) contains requirements for the developed fire alarm systems, including the number of fire sensors and its spatial arrangement. This Code includes requirements for warning systems used in the event of chemical and nuclear emergencies (NFPA 72 Standard, 2019).

A diagram of the placement of sensors over a large area, which depends on the type of sensors and the measured parameter, is presented in Fig. 3.4.

When placing sensors for analyzing the state of an infrastructure object, the problem arises of choosing the required number of sensors to obtain reliable information about the state of the infrastructure object. Fig. 3.4 shows two ways of placing sensors. In the first case, the sensors do not have overlapping zones available for measurement; in the second case, the areas can overlap. The choice of placement method is determined by the type of infrastructure facility and the type of sensor.

The layout in Fig. 3.4 includes four lines for transmitting information from sensors, with four sensors each. These lines are connected to the fire alarm system. A feature of such a connection is the need to ensure high reliability of sensors, so in the event of failure of one of them, continuity of monitoring of the state of the infrastructure object is ensured.

3.3 Analog sensors and digital sensors

The sensors used to measure the parameters of technological processes form output values that determine the features of data processing (Ida, 2014).

Let us consider some of them further. One of the important characteristics is the transfer function. The transfer function is a characteristic input and output function that determines the relationship between the output and input of the device. We can represent it by a mathematical equation or in the form of a graph with a given range of inputs and outputs. Usually, the transfer function is described by a nonlinear function $O = F(I)$, where I is the input value, O is the output value, and F is the transfer function.

There are some other characteristics that are important when we connect the sensor to the sensor network: input and output impedances of the device. The ratio of the rated voltage and the resulting current through the input port of the device with the output port open (no load) is the *input impedance*. The ratio of the rated output voltage and the short circuit current of the port (i.e., the current when the output is shorted) is the *output impedance*. When a sensor is included in a measuring system, its impedance must be taken into account.

The notion of a range of measurement of a sensor includes the lower and upper limit operating values of the device. The resolution of a sensor is the value of a minimum input signal change to which it can respond. The error involved in measurement result defines the accuracy of the device. This error is the difference between the measured and actual value.

When we measure dynamic signals, we need to know some specific characteristics of our sensors. *Frequency response* or frequency transfer function of a sensor indicates the ability of the sensor to respond to a sinusoidal input signal. The phase of the output signal can also be given. The pair amplitude gain-phase response is known as the Bode diagram. The frequency response of the device is related to the delay time, i.e., the time needed for the output to reach a steady state for a step change of input signal. The response time of the sensor is due to the inertia of the device. *Calibration* is the experimental determination of the transfer function of a sensor or actuator. One of the most important characteristics of sensors is their *reliability*. Reliability is a statistical measure that indicates the ability of a device to perform its declared function under normal operating conditions without failures for a set period of time or number of cycles. Reliability is associated with the failure rate, which determines the number of failures of the sensor for a given period of time, usually per hour.

At the output stage of an analog sensor, the output signal is continuous and proportional to the measured quantity of the physical quantity, whereas the output of a digital sensor is a binary code signal that reflects the value of the measured value in discrete quantities.

In the first case, the sensitivity of the sensor is determined by the level of the continuous signal and the transfer function. The signal must be converted to digital representation using an analog-to-digital converter.

In the second case, the sampling levels of this signal are added to the sensitivity of the analog part. When entering information into a computer, we use binary code, and the accuracy and sensitivity is determined by the long bit grid of the machine word in bits: 16 bit, 32 bit, 64 bit, etc. Next, we consider examples of analog and digital sensors.

3.3.1 Temperature sensors

Different types of sensors are used depending on the temperature range to be measured. Thermocouples are used to measure high temperatures. The thermocouple includes the connection of two dissimilar metals; at the junction, when the temperature changes, the electrical properties change, which allows us to estimate the temperature value.

Fig. 3.5 graphically shows the transfer function of a thermocouple $F(V)$ and the set of linear functions $L=(L_1(\mathrm{V}), L_2(\mathrm{V}), L_3(\mathrm{V}))$ for approximating $F(V)$. Note that since the function is nonlinear, a correction of the thermocouple readings is necessary. In digital sensors, this characteristic can be approximated based on a piecewise-linear approximation, i.e., using many linear functions. For a correct approximation, one should pay attention to points belonging to both linear straight lines.

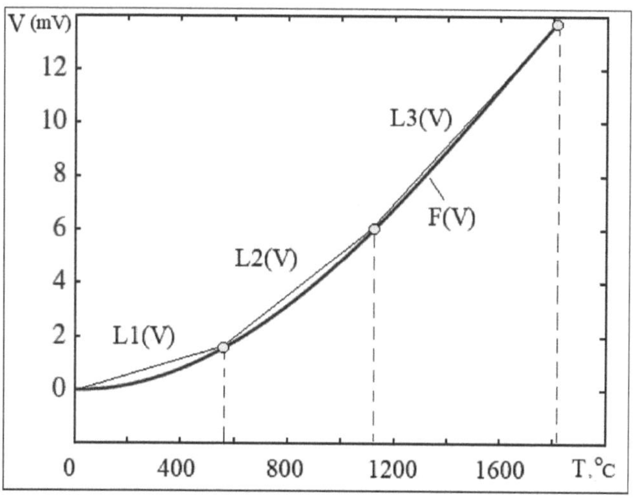

Fig. 3.5 Thermocouple conversion function.

Infrared sensors are used to measure at a distance the thermal radiation of a technological unit or technological process, and these measurements can be applied for control of large-scale infrastructure temperature state. The thermal radiation sensors can be divided into two classes: passive infrared sensors and active far infrared sensors.

In a passive infrared sensor, radiation is absorbed and converted to heat. The temperature increase is measured by a sensing element, so we can calculate radiative power.

In an active sensor, the device is heated from a power source and the variations in this power due to radiation of a process give an indication of the radiation and the temperature.

3.3.2 Image sensors

Up-to-date information on the state of technological equipment, recording information on compliance with safety regulations, and the state of the environment is carried out using image sensors. In our case, the sensor includes a lens, a sensor matrix, a processor for processing sensor matrix data, a data transmission channel, and an information storage system. The image acquisition process involves fixing the light flux reflected from the object using a sensor matrix. The sensor matrix is a microelectronic device with many miniature luminous flux detectors that respond to the intensity of the luminous flux.

Fig. 3.6 presents the image formation procedure in a simplified form. Through the lens of the system, the reflected light is directed to the sensor matrix, which contains 16 sensing elements. Each sensing element usually includes three different sensors for the *blue part* of the spectrum, the *green part* of the spectrum, and the *red part.* Using this sensor matrix, a color image is formed.

To obtain a high-quality image, matrices with a large number of sensors are used. Video sensors capture the image at certain intervals, which are stored in the internal memory and transferred for storage to an external medium. Fig. 3.7 shows the process of imaging a dynamic object.

A change in the position of an object is recorded using a video sensor and stored in a database.

3.4 Smart sensors

Advances in microelectronics make it possible to combine a measurement unit, an information processing unit (microprocessor), and a data transfer

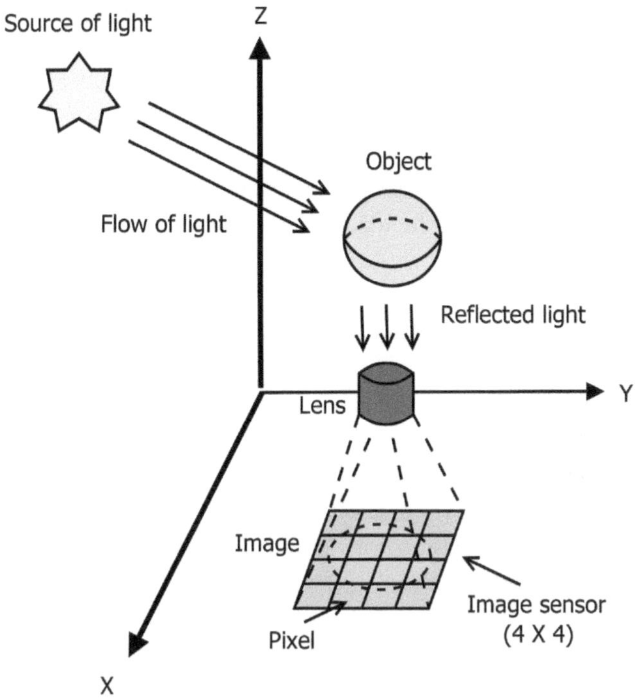

Fig. 3.6 Image processing.

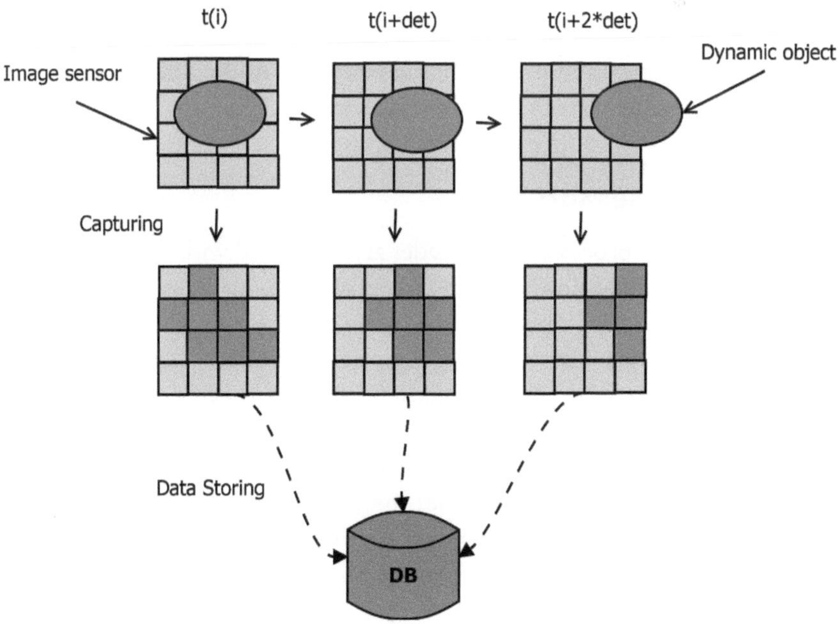

Fig. 3.7 Image data acquisition procedure.

unit in one device. Such devices are called smart sensors. They directly convert the analog signal to a digital signal, solving the problems of preprocessing a digital signal: correction, digital filtering, etc. (Huddleston & Huddleston, 2006).

In this way:

- The smart sensor takes care of the processing and monitoring of the sensor signal, reducing the load on the central control system, ensuring faster system operation.
- Intelligent sensors use a common serial bus, eliminating the need for separate wires for all sensors, significantly reducing wiring costs, large cable channels, and confusing pin assignments during maintenance or upgrades (especially if pin markers are missing or incorrectly positioned).
- Intelligent sensors have a subsystem of internal diagnostics, which reduces the cost of commissioning and start-up, as well as maintenance.
- Direct digital control of the sensor provides high accuracy that cannot be achieved using analog control systems and central processing.
- Using standard software tools allows you to add new devices to the sensor network on a "plug and play" basis.
- Individual controllers can monitor and control more than one process variable.
- The settings and calibration of the smart sensor can be changed from the central control computer.
- The cost of smart sensor systems is currently higher than that of traditional systems, but if you take into account the cost of maintenance, ease of programming, and ease of adding new sensors, then in the long term the cost of smart sensor systems will pay off.

3.5 Software sensors

The technological processes we consider are associated with the functioning of sensors under conditions of high temperatures, high pressure, aggressive chemical environment, vibration, and pressure pulsation, which can lead to failure of the sensors. The process of replacing the sensor takes time, which affects the efficiency of the technological process. In some cases, this requires a shutdown of the process.

Some qualitative characteristics of technological processes cannot be measured in real time—for example, the assessment of the quality of the products obtained during distillation. To solve the measurement problems in the cases considered, it is possible to use mathematical models of changing the values of the parameters of technological processes implemented in the form of computer programs. Thus, it is possible to evaluate the measured parameter based on the calculations. Such programs are commonly called soft sensors (Kadlec, Gabrys, & Strandt, 2009).

The classification of software sensors is presented in Fig. 3.8. Depending on the task being solved, software sensors can be divided into two classes. The model-driven sensors are based on the mathematical structures with known description such as analytical equations (the transfer function is presented as a "white box"). The data-driven sensors are based on statistical data application for transfer function defining with the help of machine learning technologies, for example, neural networks, support vector machines etc. (the model of the transfer function is presented as a "black box").

In the first case, element-wise mathematical models obtained on the basis of the description of physical and chemical processes are used to determine the output parameters. It should be noted that the development of this kind model is quite an expensive procedure. These models are used at the design stages and do not take into account the features of a particular implementation of a device, system or process. To refine the model in this case, it is necessary to solve the problems of parametric identification. In some cases, simplified mathematical models are used, obtained on the basis of simplification of the element-wise model and approximation of changes in the coefficients of equations based on various approximation methods.

In the second case, special mathematical methods are used in the development to configure linear or nonlinear approximators using textbooks and learning algorithms (Hornik, 1991). The textbook contains two data arrays: the data that are input to the model being trained and the model output. Next, an approximator is selected, for example, a neural network, and

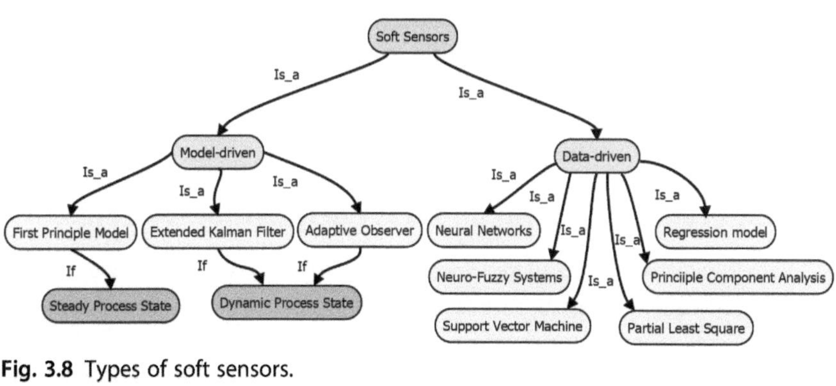

Fig. 3.8 Types of soft sensors.

the model structure is selected based on training algorithms (tuning of the approximator coefficients) (Gonzaga, Meleiro, Kiang, & Filho, 2009). Then the learning process is performed until the desired accuracy is reached. The main advantage of such models is their versatility and the possibility of adjustment based on the current data array.

3.6 Sensor fusion

To determine the state of the technological process in the chemical industry, sufficiently accurate and repeated measurements of product quality indicators in real time are necessary. Three approaches are used to obtain reliable information, based on the use of sensors for measuring product quality, determining product quality based on a laboratory analyzes performed manually and using software sensors. By combining all the information obtained on the basis of these measurement methods, more reliable information on product quality can be gathered (Liu, Srinivasan, & SelvaGuruc, 2008).

This approach is based on the idea of aggregation or date fusion. The effect is achieved by using the advantages of high-speed real-time measurements and high accuracy of analysis results in a laboratory, as well as using software sensors to compensate for missing measurements and the possibility of using statistical information within the framework of big data. Sensors for measuring temperature, flow, level, pressure, and other technological parameters are used in industrial processing plants to solve management and control problems. When analyzing the composition of products and molecular weight distribution, rather expensive devices are used, for example, chromatographs that require quality service. Therefore, the definition of product quality based on offline laboratory analyses are applied. This allows you to obtain the required information with high accuracy, but this method of obtaining information is not suitable for monitoring, control, and optimization of processes in real time due to the low sampling frequency. For this reason, soft sensors can provide missing information to predict real-time changes in difficult-to-measure quality variables.

3.7 Supervisory control and data acquisition systems

We previously examined in sufficient detail a multilevel risk management system in complex infrastructure facilities. We will now look in more detail at the features of the life cycle of data generated at various levels of the hierarchical management system. We are interested in the data received from the

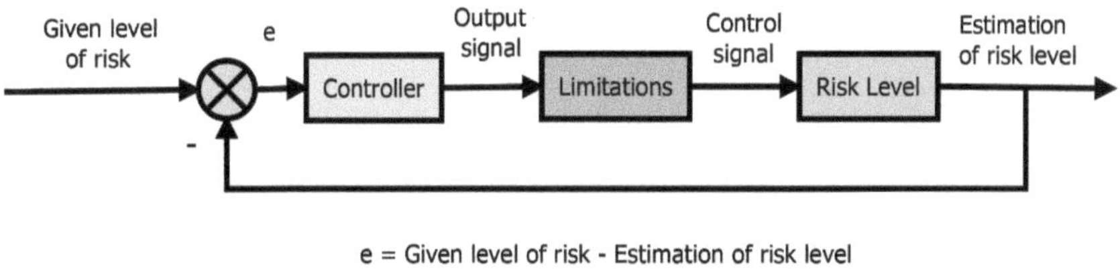

Fig. 3.9 Feedback control system.

sensors about the state of the control object, the processing of this data, and the data used to manage risks.

In the simplest case, the control error e is determined in the control system, which is determined as a result of comparing the specified risk level and assessing the current risk level obtained by analyzing the sensor data at the output of the control system (see Fig. 3.9). Further, the task of the control system is to find a control action that will reduce the amount of risk. It should be noted that restrictions on the magnitude of the control action are introduced into the system. This is due to the need to ensure the stability of control processes.

In control systems with a feedback loop, controlled variables, such as temperature or process pressure, are measured by sensors, and then the measured data is compared with the required parameters. The error measurement is sent to the input controller, at the output of which a control signal is generated. The control signal is fed to the input of the actuator, which affects a change in the controlled variable.

If our task is to control risk, we can apply the feedback control system idea. In Fig. 3.10, a risk control system is presented. A peculiarity of this

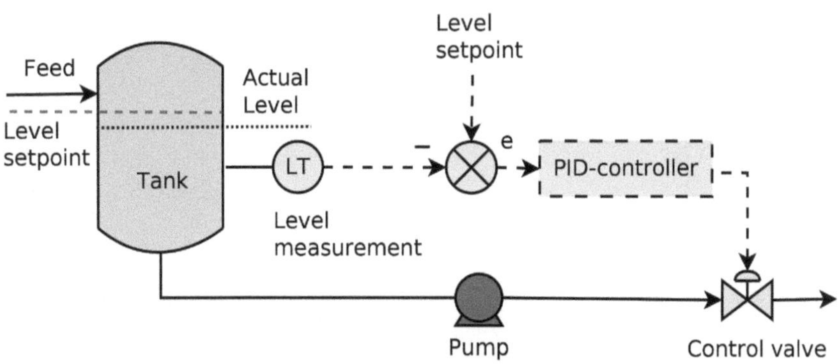

Fig. 3.10 Process feedback control system with PID controller.

system is that we need to estimate the change of risk level after applying the control system output signal for control of technological process. We must remember that there are different limitations to controlling risk level.

It is common for control of the dynamic processes to use PID controllers (see Fig. 3.10). They can provide the given dynamic of a technological process, tuned at the design stage.

An *industrial control system* (ICS) includes several types of control system:

- supervisory control and data acquisition (SCADA) system;
- distributed control system (DCS); and
- programmable logic controller (PLC).

The classification tree is presented in Fig. 3.11.

PLC-based control systems are widely used as reliable and low-cost solutions for simple control tasks. Distributed control systems are focused on classes of objects with a distributed homogeneous structure with a single control goal. For example, a control system for the distribution of temperature at predetermined points on the surface of a technical object according to a given program.

SCADA systems are used in process control in petrochemical production and belong to the class of human-machine systems. Their defining features are hierarchical architecture, the use of PLCs, the active participation of personnel in the processes of process control, and the use of network technologies and databases for the collection, storage, and transmission of data (Bailey & Wright, 2003; Clarke et al., 2003).

The architecture of a SCADA system is shown in Fig. 3.12. The system has three hierarchical levels: executive level, coordination level, and data storage level. Active elements of the coordination level are SCADA

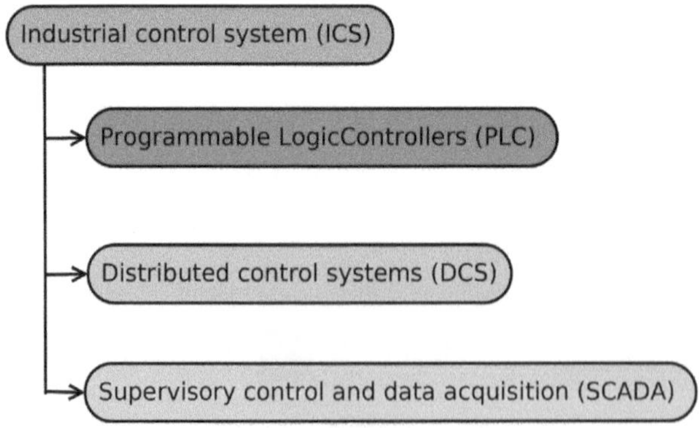

Fig. 3.11 Industrial control system classification.

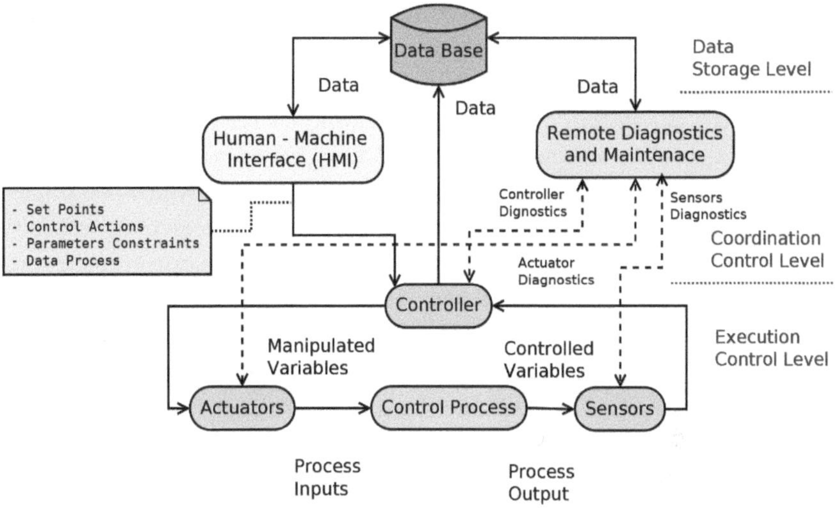

Fig. 3.12 SCADA industrial control system.

operators, who monitor and control technological processes using human-machine interfaces. A human-machine interface is a set of tools for visualizing the state of technological processes and means for entering data and commands for implementing changes to control programs if necessary. This level also includes means for remote diagnostics and maintenance of the control system using multiple network protocols for computer network elements. System diagnostic and maintenance procedures are used to prevent, detect, and recover from failures in the control system.

The data storage layer includes means for collecting, transmitting, and storing data. Data is transmitted through various communication channels including radio channels and computer networks. The collected data is stored in databases: a database for servicing operator requests and a real-time process display system and a database that stores the process status history. The latter database allows you to analyze the state of the control object at a given time interval.

3.7.1 Human machine interface

The SCADA system operator that monitors the status of process control procedures solves problems in real time and, if necessary, makes changes to the program of technological processes, controlling the entire technological chain of product production. In the event of failures, it takes the necessary measures: changing the settings, disabling the flow of components used in the technological process, etc.

To interact with the control system, human-machine interfaces are used, implemented as a system of icons on the monitor screen. During the operator's work, his/her actions are recorded using video cameras and storage systems for data entry procedures. Figs. 3.13 and 3.14 show possible implementations of the interface for monitoring the control process of the capacity filling system shown in Fig. 3.10. The interface allows the operator to monitor the fluid flow rate, the fluid level in the tank, the preset possible fluid level, and the status of the pump and valve. In Fig. 3.13, the pump and valve are turned off and are shown in *red (dark gray in print version)*. In Fig. 3.14, the fluid level has reached the permitted maximum value and the pump and

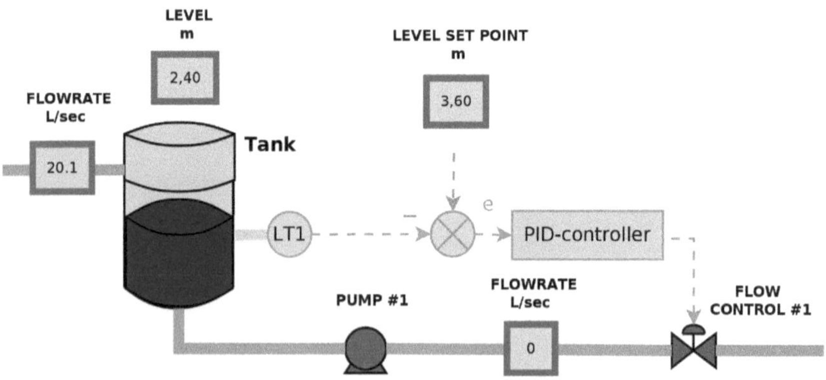

Fig. 3.13 SCADA screen for operation mode "off."

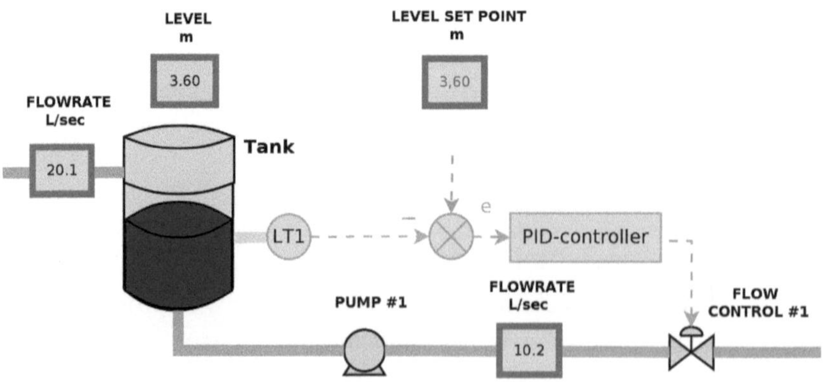

Fig. 3.14 SCADA screen for operation mode "on."

valve are automatically turned on. The status of the actuators turned on is shown in *green (light gray in print version).*

The development of the interface is given special attention, since the operator's work requires increased attention and the user-friendly interface makes it more likely to reduce risks in case of emergency situations.

3.7.2 Network architecture of SCADA system

The hierarchical SCADA enterprise system includes various levels of data collection, transmission, and storage. The generalized architecture of a multilevel system for collecting, transmitting, and storing data is presented in Fig. 3.15.

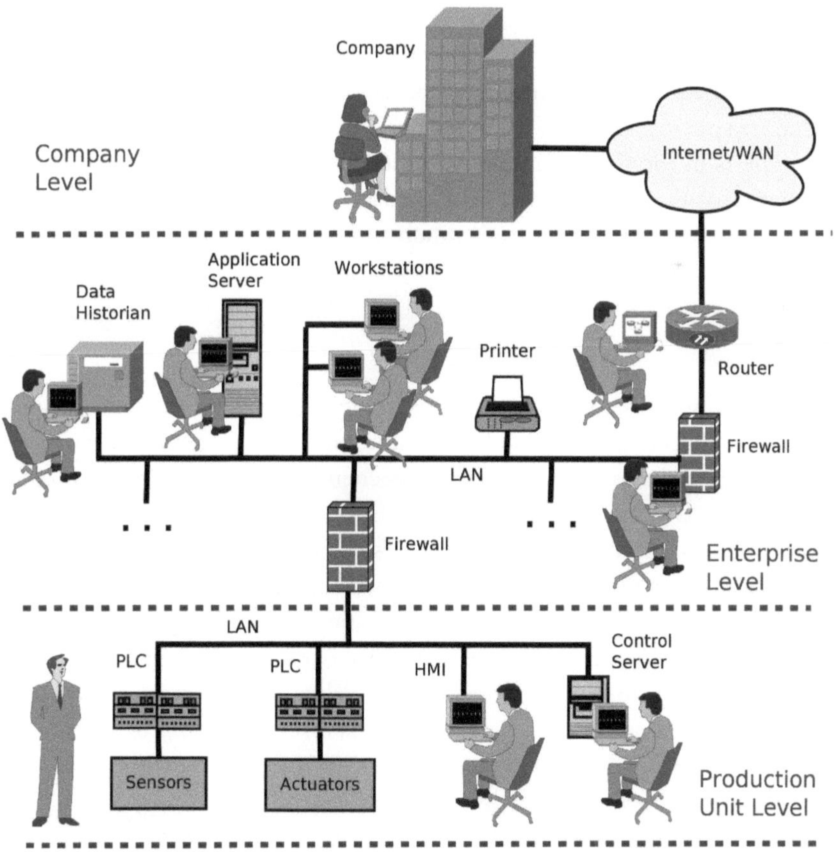

Fig. 3.15 Multilevel data acquisition system.

At the executive level of the computer network in question, all devices are connected using a local area network (LAN). PLCs, human-machine interface systems, and a server for solving the problems of integrating all devices and data storage are connected to a network at this level, on which sensors, actuators, operators, and server administrators work. They solve very important problems of control and management of technological processes and monitor the operability of the control system.

The next level of the network hierarchy is the local network of the enterprise. Using network equipment, networks of production sites are connected to it. To ensure information security, software systems—"firewalls"—are used. The local network of the enterprise makes it possible to aggregate data coming from departments and control programs for the implementation of interrelated technological processes. The history of data is stored in a data warehouse (data historian). If necessary, managers can make decisions at this level using preliminary data analysis results. This layer provides communication with the global internet/WAN network through the enterprise's firewall systems and special devices routers. Maintenance of the entire system is assigned to the system administrator, technical personnel, and experts in the field of information security.

Note that Fig. 3.15 shows a simplified architecture of the network infrastructure; in fact, this structure is much more complicated.

3.8 Measurements and big data

As noted earlier, large flows of data may be generated at an enterprise, which need to be collected, transmitted via communication channels, and stored on servers and databases. At a modern petrochemical enterprise, up to 30,000 sensors and many different SCADA systems can be used to solve management and monitoring tasks. Let us further consider how data arrays are formed, which can then be processed when they are extracted from storage systems.

When forming data arrays in an information collection system, it is necessary that each sensor has a unique label. That is, each sensor can have its own number, type, measurement range, current value, measurement error, etc.

To record information received from each sensor during the measurement, a data structure is stored in a table or database. For example, a recording format may have the following structure:

Format of recording data from the sensor = {*sensor label, measurement date, measurement time, measured parameter, readings, error*}.

In fire safety systems, sets of various sensors are used: heat sensors, fire sensors, smoke sensors. Such sensors belong to the class of alarm sensors. They give a signal, if the specified threshold is exceeded in the environmental parameter they control. These sensors are placed in accordance with regulatory requirements, and their number is determined in accordance with the size of the infrastructure facility (CFHA 72).

Consider the following example: the formation of an array of data (Valeev & Kondratyeva, 2015; Valeev, Taimurzin, & Kondratyeva, 2013).

Let there be a given point $P(x_k, y_k, z_k)$ in a large-scale infrastructure and we need to measure some parameter in it. So, we need to place some sensors into this point if we want to know the actual state of an information field in it. Our sensor can give a reliable measurement for the average meaning of a space around this point $P(x_k, y_k, z_k)$ with a radius R (see Fig. 3.16).

For example, if we have $10 \times 10 \times 10 = 1000$ points of some 3D volume of the infrastructure, we need 1000 sensors for estimation for one given parameter of the information field. In practice, we make measurements for several points only with one sensor. It is cheaper, but we measure the average meaning of our parameter and cannot control the quick change of parameters in all these points in an emergency.

Let's look at another example. Let there be some kind of infrastructure space, and we need to cover all this space with fire sensors that measure air temperature in a given range (see Fig. 3.16). In the example, 16 temperature sensors are used for monitoring.

Here we have four ranges of measured temperature set. This allows us to evaluate four levels of risk. A sensor failure signal (the sensor labeled S44) is

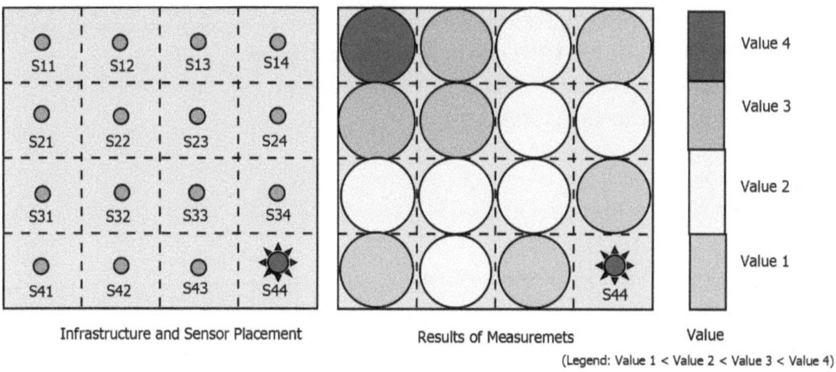

Fig. 3.16 Approximation of parameter field by the set of sensors.

sent to the information collection system. During troubleshooting, we do not have any current information on this part of the infrastructure.

From this example it follows that for a more accurate and more reliable measurement, we need to use a large number of sensors, and this in turn increases the flow of data that must be transmitted and stored. Another problem is related to the reliability of the measuring system, which includes more sensors and communication lines.

If we need to measure several parameters at a given point in the infrastructure, for example, temperature, humidity, gas concentration, smoke level, visibility, etc., we must have a set of different sensors. Usually, they have a different level of accuracy, time delay, temperature range of functioning, and so on.

Example 3.1

We can apply the example of the measurement system shown in Fig. 3.16 to estimate the amount of possible data flow. So, in our measuring system there are 16 sensors with measurements twice per second. The frequency of measurements depends on the rate of change of physical parameters. In our case, the frequency of the input signal is equal to a frequency of 0.5 Hz. Measurements depend on the frequency of changes in the parameters of the controlled process.

Below, we can do the following calculation:

16 × 2 = 32 values in 1 s.

32 × 60 = 1920 values in 1 min.

1960 × 60 = 117,600 values in 1 h.

117,600 × 24 = 2,822,400 values in 1 day.

2,822,400 × 365 = 1,030,176,000 values in 1 year.

If we have measurements for five different parameters at 16 points of infrastructure, then we will have:

1,030,176,000 × 5 = 5,150,880,000 values in 1 year.

If we have data for 10 years, then:

5,150,880,000 × 10 = 50,150,880,000 values in 10 years.

If we need 4 bytes (1 byte = 16 bits) for one measurement, then we will have the volume of data 200,150,880,000 bytes.

Note that:

1 kb = 1024 bytes

1 Mb = 1,048,576 bytes.

1 Gb = 1,073,741,824 bytes.

1 TB = 1,099,511,627,776 bytes.

Therefore:

200,150,880,000/1,099,511,627,776 = 0.182 TB.

But when we have 30,000 sensors, we need to store rather more data, as you can calculate yourself.

Example 3.2

If an infrared camera with a resolution of $640 \times 480 = 307{,}200$ pixels is used as a sensor for analyzing the thermal state of a technological installation, then we need 16 bits (2 bytes) for 1 pixel (dot on the screen) or $307{,}200 \times 2 = 614{,}400$ bytes to store the image for 16 levels of gray color.

With a frame rate of 60 Hz—that is, 60 times per second—if the formation of a new frame is performed, it will take $614{,}400 \times 60 = 18{,}864{,}000$ bytes = 18 MB of memory; so, 18 MB per sensor in 1 min.

If a video camera with a resolution of 1280×720 (30 frames per second) is used as a sensor, then for the H.264 format we need about 4 GB in 90 min, i.e., 4,073,741,824 bytes/1.5 h = 2,715,827,882 bytes per hour; $2{,}715{,}827{,}882 \times 25 \times 365 = 22.5$ TB per year.

For 16 cameras per year, the data volume is $16 \times 22.5 = 360.6$ TB. Thus, the volume of stored information increases significantly.

Input and output data generated in the process of technological operations must have unique names or labels. When developing control

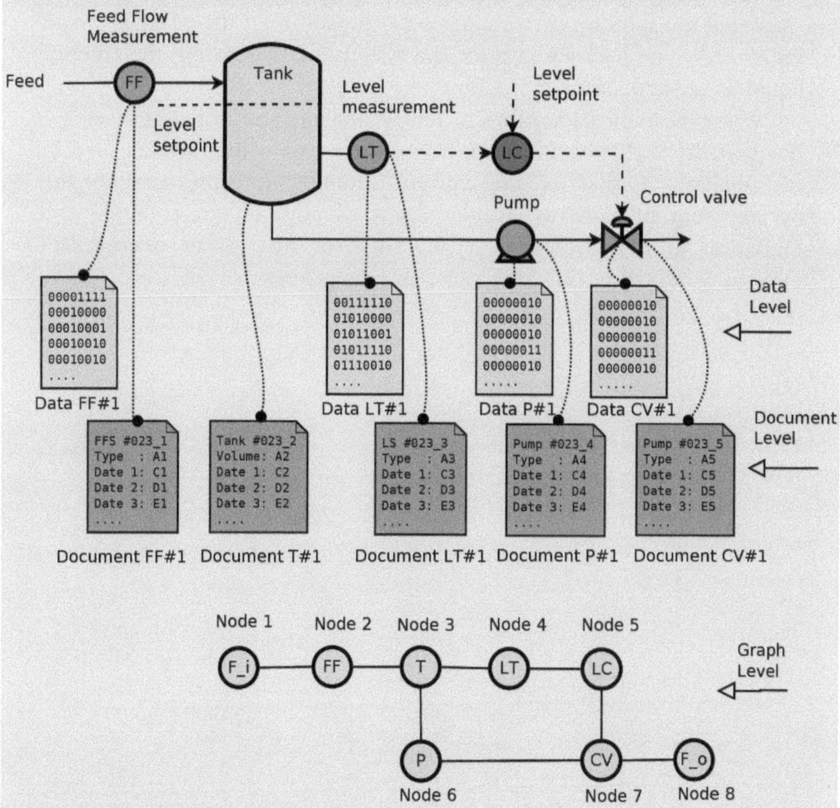

Fig. 3.17 Data flow and data structures.

programs, each software variable also requires a unique name or identifier. These identifiers, or tags, as they are usually called, are usually alphanumeric strings of characters organized in a name according to certain rules. The general agreement on the formation of names for all tags, regardless of where they are used, carries additional information. This information can be used by developers of process control programs and in solving analytic tasks.

The system under consideration, a diagram of which is presented in Fig. 3.17, includes eight elements: F_i—input fluid flow (physical quantity); FF—feed flow (physical quantity); T—tank (technical object); LT—level of liquid in the tank (physical quantity); LC—level set point (software variable of a program for a controller) (number); P—pump (technical object); CV—control valve (technical object); and F_0—output liquid flow (physical quantity).

The nodes of the graph describing the control system represent various concepts: physical quantities, technical objects, and software variables. When forming tags, recommendations are used.

Using the rules for creating labels for signals using the table shown in Fig. 3.18, we will create tags for the system for monitoring the level of liquid in the tank (see Fig. 3.17).

When forming data arrays, a rather difficult task arises of assigning unique labels to the data that we have. These data will be further used in the analysis of risks at the current moment in time and in the development of forecasts. In the event of possible errors during the formation of the "label-data" pair, an error may occur during data processing, which in turn may cause an emergency.

–	FIRST LETTER		SUCCEEDING-LETTERS		
–	MEASURED OR INITIATING VARIABLE	MODIFIER	READOUT OR PASSIVE FUNCTION	OUTPUT FUNCTION	MODIFIER
A	Analysis	–	Alarm	–	–
B	Burner, Combustion	–	User's Choice	User's Choice	User's Choice
C	Conductivity	–	–	Control	Close
D	Density/Sp. Gravity	Differential	Deviation	–	–
E	Voltage	–	Sensor	–	–
ESD	Emergency Shutdown	–	–	–	–
F	Flow Rate	Ratio (Fraction)	–	–	–
G	Gaging	–	Sight Glass, Viewing Device	–	–
H	Hand (manual)	–	–	–	High
HH	–	–	–	–	High High
I	Current (Electrical)	–	Indicate	–	–
J	Power	Scan	–	–	–
K	Time, Time Schedule	Time Rate of	–	Control Station	–
L	Level	–	Light Pilot	–	Low
LL	–	–	–	–	Low Low
M	Moisture	Momentary	–	–	Middle, Intermediate
N	User's Choice	–	User's Choice	User's Choice	User's Choice
O	User's Choice	–	Orifice, Restriction	–	Open
P	Pressure, Vacuum	–	Point (Test) Connection	–	–
Q	Quantity/Event	Integrate, Totalize	–	–	–
R	Radiation	Ratio	Record or Print	–	–
S	Speed, Frequency	Safety	–	Switch	–
T	Temperature	–	–	Transmit	–
U	Multivariable	–	Multifunction	Multifunction	Multifunction
V	Viscosity, Vibration, Mechanical Analysis	–	–	Valve, Damper, Louver	–
W	Weight, Force	–	Well or pocket	–	–
X	Unclassified	X Axis	Unclassified	Unclassified	Unclassified
Y	Event, State or Presence	Y Axis	–	Relay, Compute, Convert	–
Z	Position, Dimension	Z Axis	–	Driver, Actuator, Unclassified Final Control Element	–

Fig. 3.18 Table for creating tag names of the signals.

3.9 Summary

The chapter examined the features of the formation of data arrays at various stages of their life. Data generators are sensors, filled-in electronic and paper documents, video cameras, software systems, operator actions, etc. A multilevel SCADA was discussed. Large flows of data may be generated at an enterprise, which need to be collected, transmitted via communication channels, and stored on servers and databases. The SCADA system operator that monitors the status of process control procedures solves problems in real time. If necessary, making changes to the program of technological processes, controlling the entire technological chain of product production. In the event of failures, it takes the necessary measures: changes the settings, disables the flow of components used in the technological process, etc. At a modern petrochemical enterprise, the large number of sensors and many different SCADA systems can be used to solve management and monitoring tasks. The features of the formation of data arrays during their preparation for storage were considered.

3.10 Definitions

Sensor is an entity that allows you to determine the state or value of a given characteristic in any investigated space, the state of which is determined by a given set of parameters that can be determined and the values of which can be measured or calculated.

Intelligent sensor is a device including a measurement unit, an information processing unit (microprocessor), and a data transfer unit.

Soft sensors are mathematical models of the effects of changing the values of parameters of technological processes, implemented in the form of computer programs.

Sensor fusion is the approach based on the idea of aggregation or data fusion. The effect is achieved by using the advantages of high-speed real-time measurements and high accuracy of analysis results in a laboratory, as well as using software sensors to compensate for missing measurements and the possibility of using statistical information within the framework of big data.

SCADA (supervisory control and data acquisition systems). These systems are used in the control of technological processes in the petrochemical industry and belong to the class of human-machine systems. Their defining characteristics are hierarchical architecture, the use of PLCs, the active participation of personnel in the processes of process control, and the use of network technologies and databases for the collection, storage and transmission of data.

HMI (human machine interface). A graphical operator interface for managing and monitoring the status of a complex technical object.

LAN (local area network). A type of organization of computer network that provides communication between sensors and actuators and computers using a network protocol for exchanging data.

References

Asadzadeh, A., Arashpour, M., Li, H., Ngo, T., Bab-Hadiashar, A., & Rashidi, A. (2020). Sensor-based safety management. *Automation in Construction, 113*. https://doi.org/10.1016/j.autcon.2020.103128, 103128.

Bailey, D., & Wright, E. (2003). *Practical SCADA for industry* (pp. xiii–xiv). Oxford: Newnes.

Chohan, R. K., & Upadhyaya, B. R. (1989). Safety and fault detection in process control systems and sensors. *Fire Safety Journal, 14*(3), 167–177. https://doi.org/10.1016/0379-7112(89)90070-2.

Clarke, G., Reynders, D., Wright, E., Clarke, G., Reynders, D., & Wright, E. (2003). Fundamentals of SCADA communications. In *Practical modern SCADA protocols* (pp. 12–62). Oxford: Newnes. https://doi.org/10.1016/B978-075065799-0/50018-8 (Chapter 2).

Gao, X., Abdul Raman, A. A., Hizaddin, H. F., & Bello, M. M. (2020). Systematic review on the implementation methodologies of inherent safety in chemical process. *Journal of Loss Prevention in the Process Industries, 65*. https://doi.org/10.1016/j.jlp.2020.104092, 104092.

Gonzaga, J. C. B., Meleiro, L. A. C., Kiang, C., & Filho, R. M. (2009). ANN-based soft-sensor for real-time process monitoring and control of an industrial polymerization process. *Computers & Chemical Engineering, 33*(1), 43–49. https://doi.org/10.1016/j.compchemeng.2008.05.019.

Hornik, K. (1991). Approximation capabilities of multilayer feedforward networks. *Neural Networks, 4*(2), 251–257. https://doi.org/10.1016/0893-6080(91)90009-T.

Huddleston, C., & Huddleston, C. (2006). What are intelligent sensors, and why should I care about them? In *Embedded technology* (pp. 1–19). Burlington: Newnes. https://doi.org/10.1016/B978-075067755-4/50004-3 (Chapter 1).

Ida, N. (2014). *Sensors, actuators, and their interfaces*. SciTech Publishing.

Kadlec, P., Gabrys, B., & Strandt, S. (2009). Data-driven soft sensors in the process industry. *Computers & Chemical Engineering*, 795–814. https://doi.org/10.1016/j.compchemeng.2008.12.012.

Keren, N., West, H. H., Rogers, W. J., Gupta, J. P., & Mannan, M. S. (2003). Use of failure rate databases and process safety performance measurements to improve process safety. *Journal of Hazardous Materials, 104*(1), 75–93. https://doi.org/10.1016/S0304-3894(03)00236-X.

Liu, J., Srinivasan, R., & SelvaGuruc, P. N. (2008). Practical challenges in developing data-driven soft sensors for quality prediction. *Computer Aided Chemical Engineering, 25*, 961–966. https://doi.org/10.1016/S1570-7946(08)80166-6.

NFPA 72 Standard. (2019). Retrieved 5 April 2020, from https://www.nfpa.org/codes-and-standards/all-codes-and-standards/list-of-codes-and-standards/detail?code=72.

Palanivelu, R., & Srinivasan, P. S. S. (2020). Safety and security measurement in industrial environment based on smart IOT technology based augmented data recognizing scheme. *Computer Communications*, *150*, 777–787. https://doi.org/10.1016/j.comcom.2019.12.013.

Valeev, S., & Kondratyeva, N. (2015, August). Technical safety system with self-organizing sensor system and fuzzy decision support system. In *Presented at the 2015 IEEE international conference on fuzzy systems (FUZZ-IEEE), Istanbul, Turkey*, IEEE. https://doi.org/10.1109/FUZZ-IEEE.2015.7337962.

Valeev, S. S., Taimurzin, M. I., & Kondratyeva, N. V. (2013). An adaptive data acquisition system in technical safety systems. *Automation and Remote Control*, *74*, 2137–2142. https://doi.org/10.1134/S0005117913120151.

Wang, Y., Zheng, G., & Wang, X. (2019). Development and application of a goaf-safety monitoring system using multi-sensor information fusion. *Tunnelling and Underground Space Technology*, *94*. https://doi.org/10.1016/j.tust.2019.103112, 103112.

Yousefi, A., & Rodriguez Hernandez, M. (2020). A novel methodology to measure safety level of a process plant using a system theory based method (STAMP). *Process Safety and Environmental Protection*, *136*, 296–309. https://doi.org/10.1016/j.psep.2020.01.035.

Zhou, X., & Peng, T. (2020). Application of multi-sensor fuzzy information fusion algorithm in industrial safety monitoring system. *Safety Science*, *122*. https://doi.org/10.1016/j.ssci.2019.104531, 104531.

CHAPTER 4

Databases and big data technologies

4.1 Data ecosystem

The task of obtaining new knowledge based on data processing is data science. From the data generation phase to its use, many different data transformations and processes are performed. As a result, data can be used at the top level of a process safety management system (Fig. 4.1).

A data ecosystem graph is presented in Fig. 4.2. This graph reflects the main stages of the data life cycle, models, and technologies that are used to form the final result.

The main concept in the information processing technologies we are considering is the concept of data. Data refers to the presentation of information in a formalized form, which allows it to be transmitted via communication channels, stored on a computer, and processed to generate new data.

Since the data is processed on computers, in which resources are limited by the amount of RAM and the type of processor, the data has an allowed format and allowed operations. These formats and operations are supported on all types of computers and are called data types. For example, an integer data type has a simple structure, is an object of a given range of integers, and valid actions include arithmetic operations on integers. Data is combined into sets or arrays of data that can be accessed, downloaded, modified, and deleted. The processing of various data is associated with the execution of permitted actions for this data type, such as arithmetic or logical operations with binary data, combining arrays of data of various types, sorting data, and operations with data in a text format: editing, sorting, combining, storing, extracting, displaying or printing.

Considering the information systems of an industrial enterprise, we use the concepts of information and data. This section focuses on data processing methods and tools, and although the information is based on data, we will further distinguish between these two concepts. Information is associated with the interpretation of available data, which allows you to make targeted decisions when managing the object, i.e., information is generated based on data.

Process Safety and Big Data
https://doi.org/10.1016/B978-0-12-822066-5.00003-0

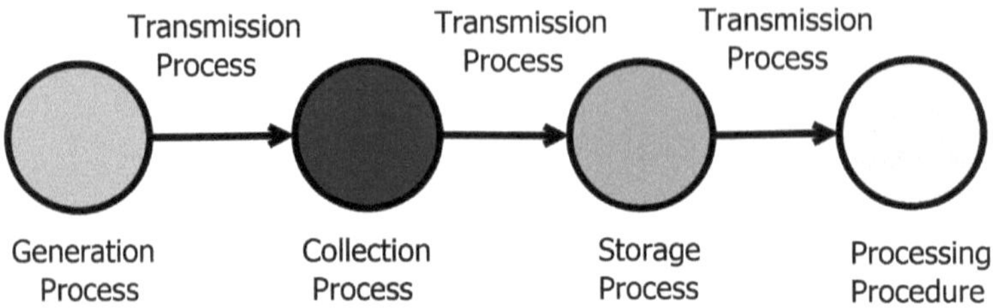

Fig. 4.1 Data life cycle.

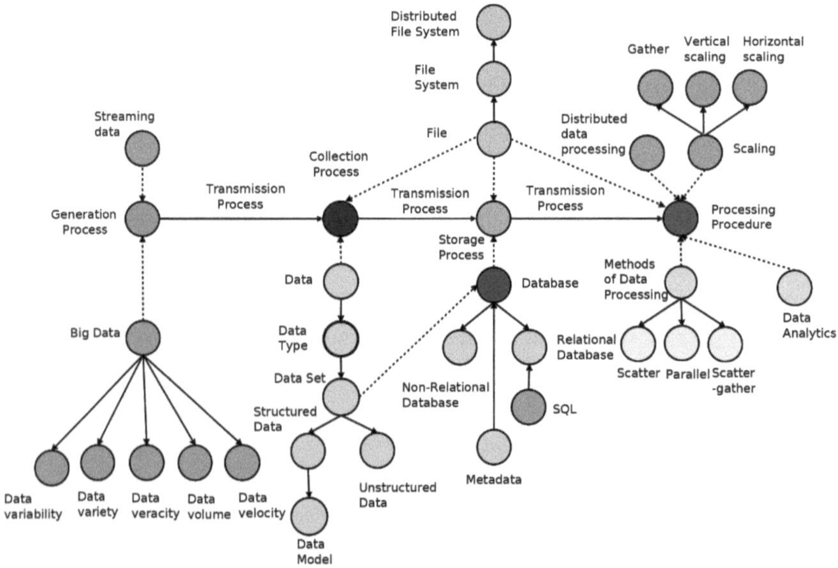

Fig. 4.2 Data ecosystem.

An information system is a complex consisting of software and data storage systems, developed as a whole and designed to automate a certain type of activity at an industrial enterprise. Information systems store data, facts, and documents. For example, the facts are data about employees of the enterprise: surname, name, position, etc.; or equipment: brand, year of manufacture, manufacturer, etc.

In other words, a fact in an information system is represented as a set of some properties, or attributes, the quantitative value of which, as a rule, is expressed by a simple data type in the form of an integer, or real number, or string.

A document, unlike a fact, cannot be expressed by a simple structure. Examples of documents include regulations, texts of orders and instructions, accounting documents, plans of an enterprise, sound recordings, videos, etc. The structure of objects that we designate as documents can be represented in any format of electronic entities:

- For text documents, TXT text format, PDF format, EPUB format, HTML format, etc. are used.
- The spreadsheet format is used for sets of interdependent data.
- For graphic images, PNG, JPEG, GIF, TIFF, etc. are used.

A document is an object of an arbitrary structure stored in an information database containing information of an arbitrary nature, access to which can be obtained by its requisites.

Requisites of the document are a set of properties of this document, allowing you to identify it uniquely. Examples of requisites include the name of the document, its number, date of creation, names of creators, electronic signature, etc.

Data is stored and processed in information systems. Information systems include the following main components (Fig. 4.3):

- user interface in the form of a menu system or command line;
- business logic, reflecting the purpose of the system;
- interface with the database;
- database; and
- intersystem interface.

When developing the system and its subsystems, various programming languages are used. The implementation of system tasks is performed using computer systems of varying complexity. The interaction of the system with the user and other systems is based on computer networks.

4.2 Algorithms and complexity

Data processing in information systems is based on the use of various algorithms. An algorithm is a description that defines the procedure for finding a solution (Guo, Du, Qi, & Qian, 2012; Valeev, Kondratyeva, & Lutov, 2019; Venkateswarlu, Jujjavarapu, Venkateswarlu, & Jujjavarapu, 2020; Witten, Frank, Hall, & Pal, 2017). The description contains a sequence of steps to perform any action from the set of allowed actions in a given sequence. Algorithms are described using the algorithms description languages in the class of formalized languages, or using various graphical tools. The implementation of algorithms in programming languages is focused on a

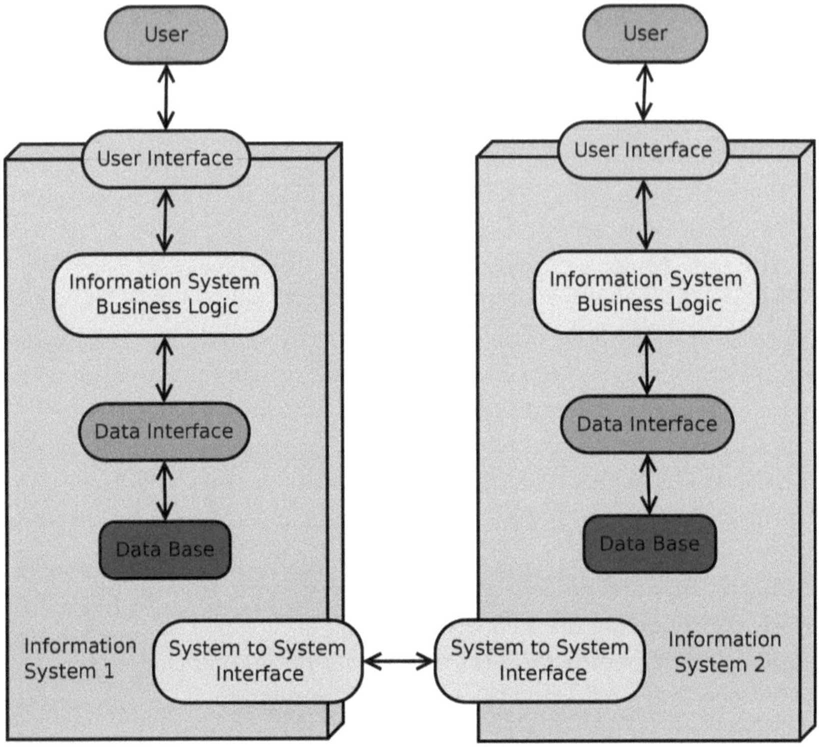

Fig. 4.3 Information system architecture.

sequence of steps using computers (Scott, 2009). The graphic form of the description of the algorithms allows us to present the main stages of obtaining a solution to a problem using graphical tools. The graphic form is used in the design of information systems by developers (Weilkiens & Oestereich, 2007).

When processing big data, it is necessary to solve the problem of searching for information and determining the numerical values of the required function. Consider an example of an algorithm for finding the largest of a given set of numbers (Fig. 4.4).

The algorithm includes two initial steps: reading and storing the first number. After that, a cycle of calculations is performed in which subsequent numbers are read and verified. The assumption is made that there is an array of data whose minimum size is $n = 2$. We rewrite the steps of the algorithm in a programming language. The syntax of the language contains generalized types of commands, and is not tied to any implementation. Further, as an

```
Algorithm
Step 1. Read the value of the first number.
Step 2. Assign a value of a read number to a variable named A
Step 3. Read the next number
Step 4. If the read number is greater than A, assign A the value of the given number.
Step 5. If there are more numbers, go to step 3.
Step 6. Print the value A.
```

```
Program MaxN(Input, Output);
Begin
ReadLn(A);                {Step 1 and 2}
Repeat
ReadLn(X);                {Step 3}
If X > A then A=X;        {Step 4}
UntilEOF;                 {Step 5}
WriteLn(A);               {Step 6}
End.
```

Fig. 4.4 Algorithm and program for finding the largest number.

example, we consider the analysis of data obtained during the measurement of three parameters p_1, p_2, p_3 that determine the state of the process. One of the tasks is sorting the data in the columns of the table in increasing order.

An example of the measurement results recorded in Table 1 is shown in Fig. 4.5. Next, the data is sorted by the algorithm (see Fig. 4.6) (Step 2–Step 4). In Step 5, as a result of the analysis, we can conclude that in the 6th row of Table 1 there may be an error in recording measurements of the parameter p_3 or a deviation of the parameter p_3 caused by other reasons.

In the event of a change in the sorting analysis of the sequence of sorted columns of Table 1, the results may look different (see Fig. 4.7). Therefore, it

Table 1

#	p1	p2	p3
1	1	2	2
2	3	2	4
3	2	3	4
4	4	3	4
5	4	3	3
6	5	4	3
7	2	3	4
8	3	2	2
9	1	2	2
10	3	2	2

Table 2

#	p1	p2	p3
1	1	2	2
9	1	2	2
3	2	3	4
7	2	3	4
8	3	2	2
10	3	2	2
2	3	2	4
5	4	3	3
4	4	3	4
6	5	4	3

Table 3

#	p1	p2	p3
1	1	2	2
9	1	2	2
8	3	2	2
10	3	2	2
2	3	2	4
3	2	3	4
7	2	3	4
5	4	3	3
4	4	3	4
6	5	4	3

Table 4

#	p1	p2	p3
1	1	2	2
9	1	2	2
8	3	2	2
10	3	2	2
5	4	3	3
6	5	4	3
3	2	3	4
7	2	3	4
2	3	2	4
4	4	3	4

Fig. 4.5 Example of data analysis (sorting order of columns p_l, p_2, p_3).

Algorithm
Step 1 Get data and write to the table
Step 2 Sort the rows in increasing order of the parameter p1
Step 3 Sort the rows in increasing order of the parameter p2
Step 4 Sort the rows in increasing order of the parameter p3
Step 5 Analyzing the obtained results

Fig. 4.6 Algorithm of data analysis.

Table 1

#	p1	p2	p3
1	1	2	2
2	3	2	4
3	2	3	4
4	4	3	4
5	4	3	3
6	5	4	3
7	2	3	4
8	3	2	2
9	1	2	2
10	3	2	2

Table 2

#	p1	p2	p3
1	1	2	2
9	1	2	2
3	2	3	4
7	2	3	4
8	3	2	2
10	3	2	2
2	3	2	4
5	4	3	3
4	4	3	4
6	5	4	3

Table 3

#	p1	p2	p3
1	1	2	2
9	1	2	2
8	3	2	2
10	3	2	2
5	4	3	3
6	5	4	3
3	2	3	4
7	2	3	4
2	3	2	4
4	4	3	4

Table 4

#	p1	p2	p3
1	1	2	2
9	1	2	2
8	3	2	2
10	3	2	2
2	3	2	4
5	4	3	3
3	2	3	4
7	2	3	4
4	4	3	4
6	5	4	3

Fig. 4.7 Example of data analysis (sorting order of columns p_l, p_3, p_2).

is necessary to develop a criterion for determining the deviation of the analyzed data from the expected. This criterion depends on the type of problem being solved.

If it is the case that the data array includes a large data set, the solution considered in the example of the analysis problem requires the use of a large proportion of the computer's random access memory (RAM) and multiprocessing.

The analysis problem can be solved using two computers. The first computer performs the analysis task for the first half of the table, and the second for the second half of the table (see Fig. 4.8). After the completion of the work of these computers, the analysis results can be transferred to a third computer, which combine the two tables. Such data processing is used in multimachine computing systems and is called parallel computing (Buyya et al., 2013; Oshana, 2016).

Given the need to perform many operations when working with data, solving a number of problems using algorithms requires an assessment of the complexity of the algorithm for a conscious choice of the analysis method. The complexity of algorithms is investigated in the theory of complexity of algorithms.

Table 1 (1)

#	p1	p2	p3
1	1	2	2
2	3	2	4
3	2	3	4
4	4	3	4
5	4	3	3

Table 2 (1)

#	p1	p2	p3
1	1	2	2
3	2	3	4
2	3	2	4
4	4	3	4
5	4	3	3

Table 3 (1)

#	p1	p2	p3
1	1	2	2
2	3	2	4
3	2	3	4
4	4	3	4
5	4	3	3

Table 4 (1)

#	p1	p2	p3
1	1	2	2
5	4	3	3
2	3	2	4
3	2	3	4
4	4	3	4

Table 1 (2)

#	p1	p2	p3
6	5	4	3
7	2	3	4
8	3	2	2
9	1	2	2
10	3	2	2

Table 2 (2)

#	p1	p2	p3
9	1	2	2
7	2	3	4
8	3	2	2
10	3	2	2
6	5	4	3

Table 3 (2)

#	p1	p2	p3
9	1	2	2
8	3	2	2
10	3	2	2
7	2	3	4
6	5	4	3

Table 4 (2)

#	p1	p2	p3
9	1	2	2
8	3	2	2
10	3	2	2
6	5	4	3
7	2	3	4

Fig. 4.8 Example of data analysis (sorting order of columns p_l, p_2, p_3) for two tables.

The analysis of many different algorithms has been performed, which allows many computational tasks to be divided into classes by complexity and, within the class, to choose the most acceptable data processing algorithm taking into account the specifics of the task and data type (Vaidyanathan et al., 2015).

4.3 Modern databases

The measurement results obtained by the sensors are transmitted via radio channels or via computer networks and stored in digital data acquisition systems. Each measured parameter characterizing the technological process has its own name or tag, for example, the temperature of the product ("Product" when writing to the information collection system), for example, "$T_g(t) = 250$"; this record includes the label of the measured parameter "T_g," measurement time "t," and its numerical value is "250."

In the future, we assume that "data" is a list structure:

$$\text{Data} = \{\text{tag}, \{\text{content}\}\},$$

where tag is a unique identifier from the given list and content is digital code (number, list of characters, etc.).

Fig. 4.9 shows the process of collecting and storing data. To unify these processes, standard data presentation formats, standard storage models, and standard methods for receiving data in response to a request from an operator or analyst have been developed.

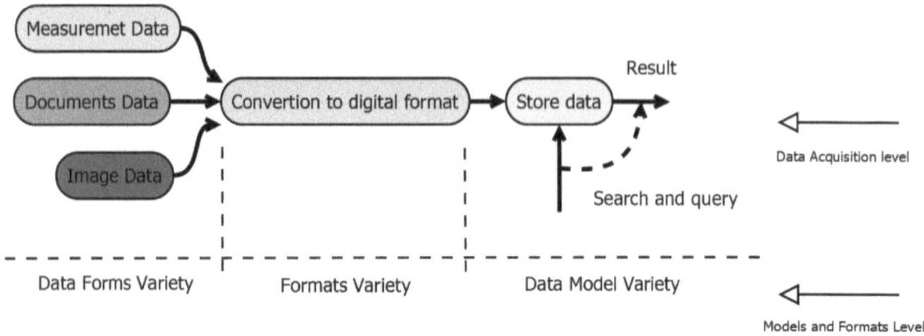

Fig. 4.9 Data acquisition level and abstract data models level.

Consider further computer systems for storing data. These systems include *databases* and *database management systems*. A database is a hardware-software system for logical collection of data, which is managed by a software database management system. The semantic graph of database system definition is presented in Fig. 4.10. A database includes the data and the indexes (logfiles) necessary for its management. The database management system is a software tool used for manipulating (storing, querying) a collection of logically related data.

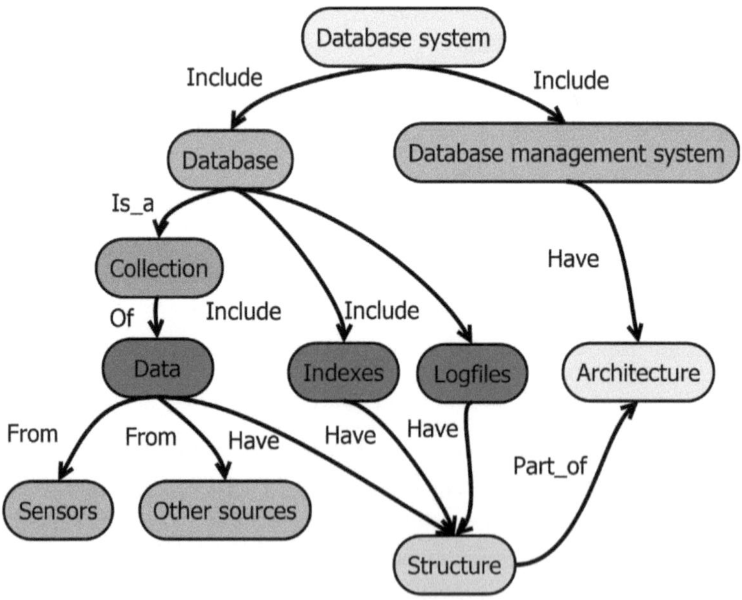

Fig. 4.10 Semantic graph of database notion definition.

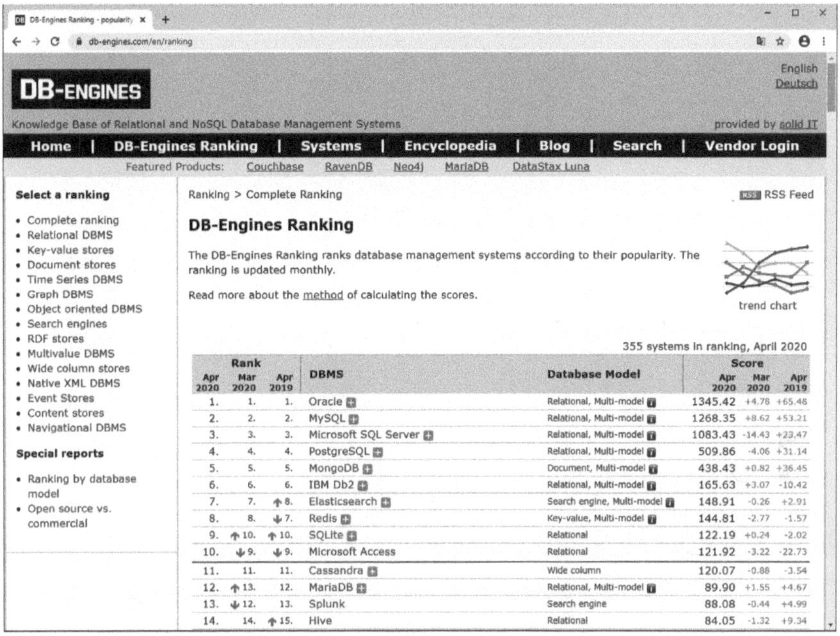

Fig. 4.11 DB-Engines ranking (https://db-engines.com/en).

More than 350 database management systems are available on the software market; an example (DB-Engines, n.d.) is given in Fig. 4.11. The classification of these databases will be discussed later.

A database management system that supports the *relational data model* is often called a relational database management system. If it supports other data models, it is known as a *NoSQL* system. When reporting incidents related to critical situations, information can be placed in a spreadsheet (see Fig. 4.12) or in a relational database (see Fig. 4.13). With large amounts of information, it is usually placed in a DBMS. The type of DBMS used is determined by developers, who take into account many different factors: cost, amount of data, history of the formation of data arrays, etc.

4.3.1 SQL and NoSQL databases

The most popular relational database management systems are those that support the relational or table-oriented data model. The model includes a set of tables or relation schemas defined by the tables' names and a fixed number of attributes with fixed data types (Hardy & Stobart, 2003).

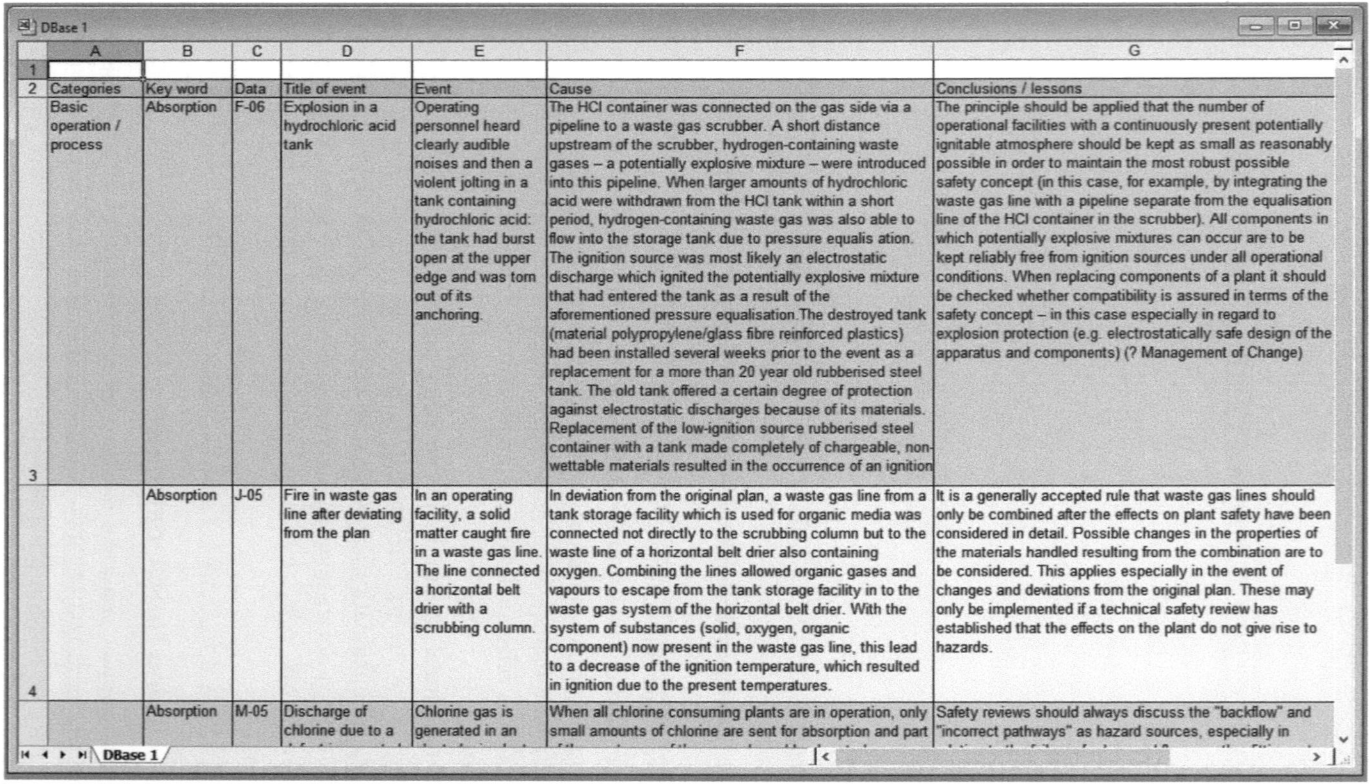

DBase 1

	A	B	C	D	E	F	G
1							
2	Categories	Key word	Data	Title of event	Event	Cause	Conclusions / lessons
3	Basic operation / process	Absorption	F-06	Explosion in a hydrochloric acid tank	Operating personnel heard clearly audible noises and then a violent jolting in a tank containing hydrochloric acid: the tank had burst open at the upper edge and was torn out of its anchoring.	The HCl container was connected on the gas side via a pipeline to a waste gas scrubber. A short distance upstream of the scrubber, hydrogen-containing waste gases – a potentially explosive mixture – were introduced into this pipeline. When larger amounts of hydrochloric acid were withdrawn from the HCl tank within a short period, hydrogen-containing waste gas was also able to flow into the storage tank due to pressure equalis ation. The ignition source was most likely an electrostatic discharge which ignited the potentially explosive mixture that had entered the tank as a result of the aforementioned pressure equalisation.The destroyed tank (material polypropylene/glass fibre reinforced plastics) had been installed several weeks prior to the event as a replacement for a more than 20 year old rubberised steel tank. The old tank offered a certain degree of protection against electrostatic discharges because of its materials. Replacement of the low-ignition source rubberised steel container with a tank made completely of chargeable, non-wettable materials resulted in the occurrence of an ignition	The principle should be applied that the number of operational facilities with a continuously present potentially ignitable atmosphere should be kept as small as reasonably possible in order to maintain the most robust possible safety concept (in this case, for example, by integrating the waste gas line with a pipeline separate from the equalisation line of the HCl container in the scrubber). All components in which potentially explosive mixtures can occur are to be kept reliably free from ignition sources under all operational conditions. When replacing components of a plant it should be checked whether compatibility is assured in terms of the safety concept – in this case especially in regard to explosion protection (e.g. electrostatically safe design of the apparatus and components) (? Management of Change)
4		Absorption	J-05	Fire in waste gas line after deviating from the plan	In an operating facility, a solid matter caught fire in a waste gas line. The line connected a horizontal belt drier with a scrubbing column.	In deviation from the original plan, a waste gas line from a tank storage facility which is used for organic media was connected not directly to the scrubbing column but to the waste line of a horizontal belt drier also containing oxygen. Combining the lines allowed organic gases and vapours to escape from the tank storage facility in to the waste gas system of the horizontal belt drier. With the system of substances (solid, oxygen, organic component) now present in the waste gas line, this lead to a decrease of the ignition temperature, which resulted in ignition due to the present temperatures.	It is a generally accepted rule that waste gas lines should only be combined after the effects on plant safety have been considered in detail. Possible changes in the properties of the materials handled resulting from the combination are to be considered. This applies especially in the event of changes and deviations from the original plan. These may only be implemented if a technical safety review has established that the effects on the plant do not give rise to hazards.
		Absorption	M-05	Discharge of chlorine due to a	Chlorine gas is generated in an	When all chlorine consuming plants are in operation, only small amounts of chlorine are sent for absorption and part	Safety reviews should always discuss the "backflow" and "incorrect pathways" as hazard sources, especially in

DBase 1

Fig. 4.12 MS Excel datasheet.

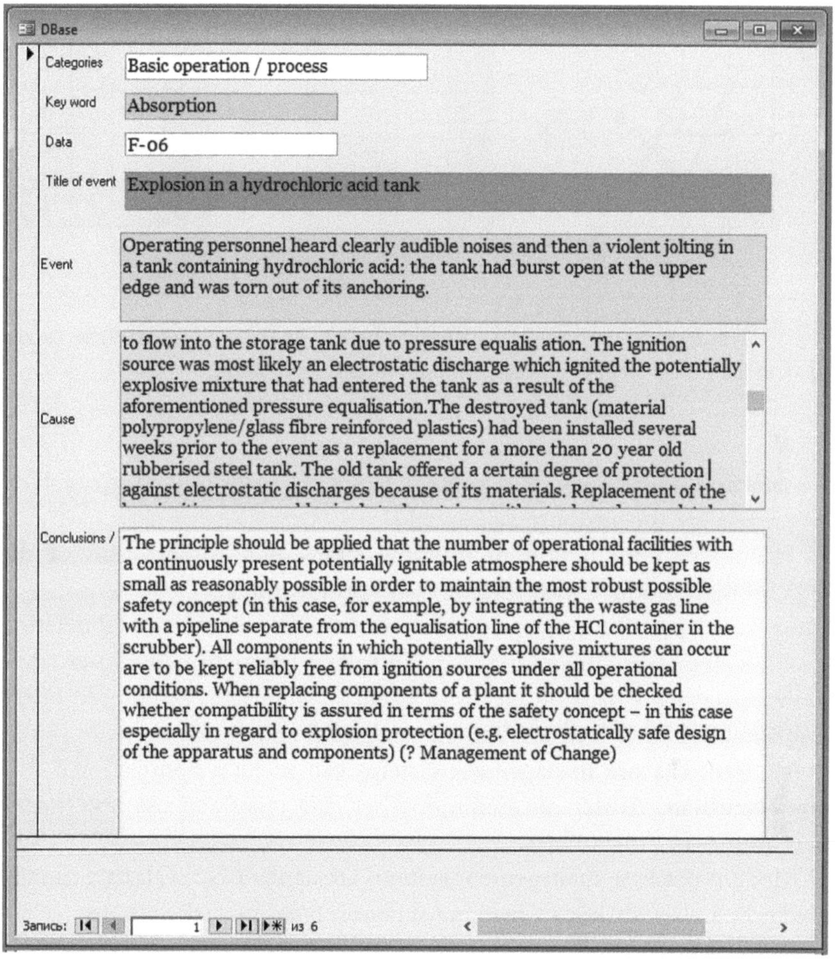

Fig. 4.13 MS Access database.

A unique record in the database corresponds to a row in the table and consists of the values of each attribute of the row. Uniqueness is ensured by the choice of key and its meaning. Row attributes are records that are located in the vertical columns of a table (see Fig. 4.14).

Thus, the database consists of a set of uniform records. The basic operations on records are union, intersection, and difference. When forming a database structure, several tables can be used that are linked through their unique links.

Basic operations applied in relational databases are:

- projection or selecting of a subset of attributes or columns of the table;

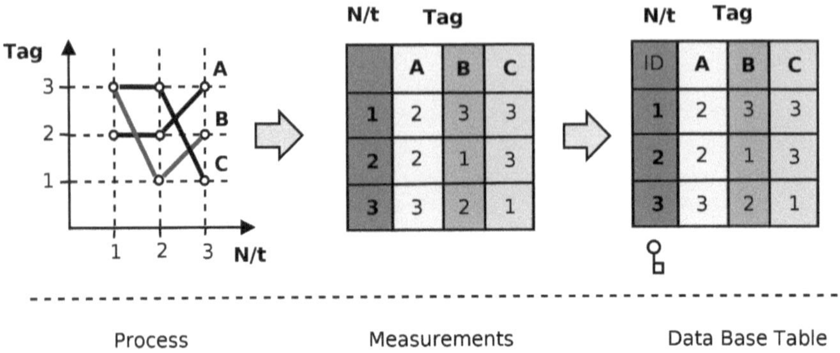

Fig. 4.14 Life cycle stages "process—measurements—database."

- selection or defining of a subset of records according to given filter criteria for the attribute values; and
- joining or conjunction of multiple tables as a combination of the Cartesian product with selection and projection.

These basic operations, operations for creation, modification, and deletion of table schemas, and operations for controlling transactions and user management are performed by means of database languages. Structured query language (SQL) is a well-established standard for such languages.

Modern database management systems can include nonrelational data type extensions. A user can define his/her data types and use inheritance mechanisms and hierarchies.

Modern database management systems are applied NoSQL data models (Celko & Celko, 2014a). There are different types of such systems:

- key-value store (KVS);
- wide column store (WCS);
- document store (DS);
- graph DBMS (GDBMS);
- RDF store (RDFS);
- native XML DBMS;
- content store (CS); and
- search engine (SE).

Depending on the given task, these database management systems can provide high performance and efficient distribution of data across various storage nodes. This provides scalability when we need to add new nodes, fault tolerance, and high flexibility, due to the use of a data model without a defined scheme.

4.3.2 Graph databases

A graph database (GDB) is a database that uses graph structures for queries and apply nodes and edges to store data sets (Celko & Celko, 2014b; McKnight & McKnight, 2014; Powell, Hopkins, Powell, & Hopkins, 2015). The main data representation model is a graph, which associates data items in a warehouse with a set of nodes and edges. Edges represent relationships between nodes. These relationships allow you to manipulate and retrieve the data in the data storage using simple operations performed on graphs. Thus, graph databases allow you to store relationships between data as the basic elements of a data array. Relationships are retrieved quickly since they are stored in the database. A lot of relationships can be represented as a graph during visualization, which is quite convenient when analyzing related data.

Graph databases are classified as NoSQL; sometimes they are allocated to a separate database class. The methods for storing graph databases depend on the implementation of the DBMS. In some databases of this type, data is stored in a table, which allows the use of classic query options. Other databases may use a key-value type. When executing queries in this type of database, various graph search algorithms are implemented. These algorithms have been well researched (Beretta et al., 2019; Dondi et al., 2019).

When solving the problems of processing big data for risk analysis in large infrastructure systems, an analysis of the state of relations between systems and elements is necessary.

In Chapter 1, we discussed the description of complex systems applying the notion of undirected graphs. This allows us to evaluate the impact of relationships between elements on the integrity of the system, and thereby assess the risks. In practice, a fairly time-consuming procedure is to assess the state of connections between elements of complex technical objects; so, in a fully connected graph, the number of connections is not a linear function.

Fig. 4.15 shows the procedure for storing the results of a risk assessment in a graph database. This procedure includes several basic steps.

4.4 Big data technologies

Information streams, including data from tens of thousands of sensors and actuators, documents, photographs, audio files, and video files require special means of transmission, storage, and processing. These tools include software systems focused on processing large volumes of various data. To ensure

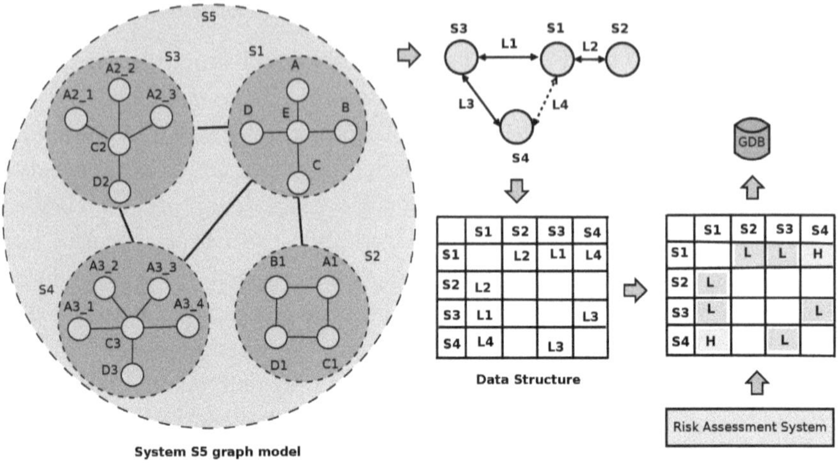

Fig. 4.15 Graph database for risk assessment.

reliable storage and scaling, specialized data storage and processing systems are used. Next, we consider software systems that allow us to solve the problems of processing big data.

4.4.1 Clusters systems

A computer cluster is a set of connected computers that solve a common data processing problem. Unlike computers connected in a network, in computer clusters, each node performs the same task under the control of software (Basford et al., 2020; López & Baydal, 2017). Cluster components are interconnected via fast local area networks. Traditionally, all nodes consist of the same type of computers controlled by the same type of operating systems. Clusters are commonly used to improve computing performance. In contrast to highly reliable powerful computer systems, clusters are cheaper to scale, i.e., increase computing power by adding new nodes. Hundreds or thousands of clusters can be used in data centers (Sterling et al., 2018a).

Within the framework of the internet of things, edge computing technologies—cluster solutions based on minicomputers—can be used. An example of such a cluster structure is shown in Fig. 4.16.

The system includes all the basic elements of a cluster: a microchip (1), a minicomputer (2), a minicomputer in a case with a monitor (3), a Wi-Fi router for connecting the cluster to the internet (4, 7), a network hub (5), a data storage server (6, 9), and power supplies (7).

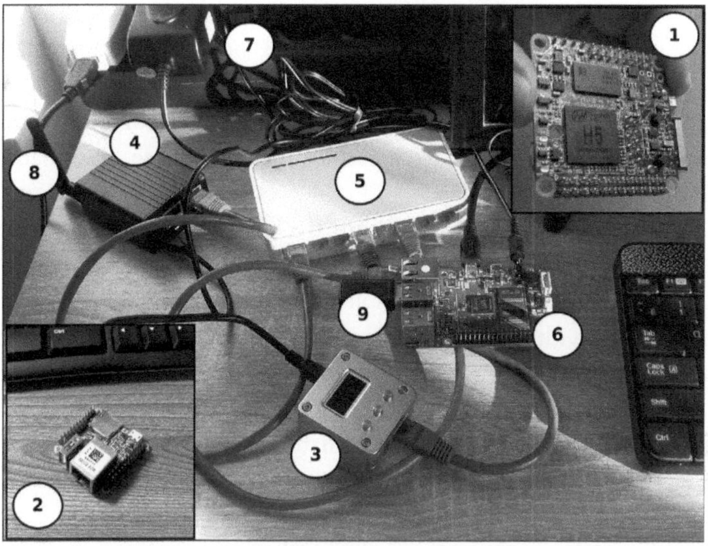

Fig. 4.16 Cluster system with minicomputers.

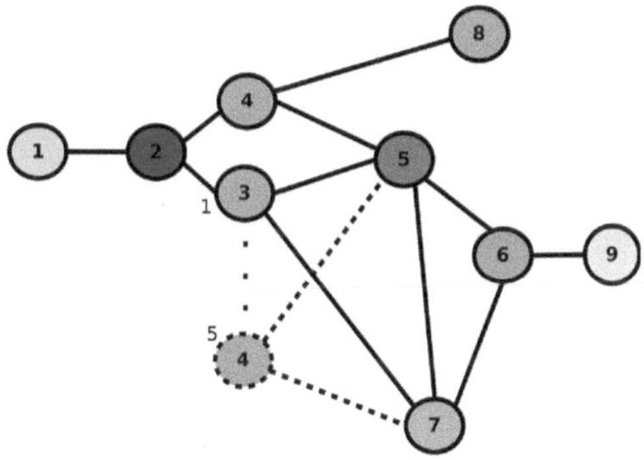

Fig. 4.17 Graph model of the cluster.

In Fig. 4.17, a cluster system is presented, where element 8 allows communication via a radio channel with a local area network (LAN), and element 9 provides radio communication with peripheral devices: a keyboard and a monitor.

This model allows us to solve the problems of risk analysis for a cluster system as an element of the hierarchical process safety management system.

4.4.2 MapReduce

MapReduce is a distributed computing model developed by Google and used in big data technologies for organizing parallel processing of large (up to several petabytes) data sets in computer cluster systems, and as a framework for solving distributed tasks on cluster nodes (Buyya et al., 2013; Sterling et al., 2018b).

The essence of MapReduce is to divide the information array into parts, carry out parallel processing of each part on a separate node, and finally combine all the results.

The execution of programs, within the framework of MapReduce, is automatically parallelized and executed on cluster nodes. At the same time, the system itself takes care of the implementation details: splitting the input data into parts, dividing tasks by cluster nodes, and processing failures and messages between distributed computers. This data processing technology is used to index web content, count words in a large file, count the frequency of accesses to a given web address, calculate the volume of all web pages of each URL address of a particular host, create a list of all addresses with the necessary data, and other processing tasks associated with huge arrays of distributed information.

The MapReduce fields of application also include distributed search and sorting of data, processing statistics of network log files, document clustering, machine learning, and statistical machine translation (Fig. 4.18).

During calculations in MapReduce, the set of input key/value pairs is converted to the set of output key/value pairs.

MapReduce uses the following basic procedures:

- Map—preliminary processing of input data specified in the form of a large list of data. The master node of the cluster receives this list, divides it into parts, and passes it to the working nodes. After that, each node applies the map function to local data and writes the result in the key-value format to temporary storage.
- Shuffle—nodes redistribute data based on the keys previously created by the map function, so that all data with the same keys is placed on one of the working nodes.
- Reduce—parallel processing by each working node of each data group according to the specified order of keys and "gluing" the results to the master node. The result obtained after passing through all the necessary data processing procedures is the solution to the original problem.

It should be noted that this technology is not applied in its pure form in streaming big data systems, where it is necessary to quickly process large amounts of continuously incoming information in real time. In practice, such tasks need to be addressed in internet of things technologies. However,

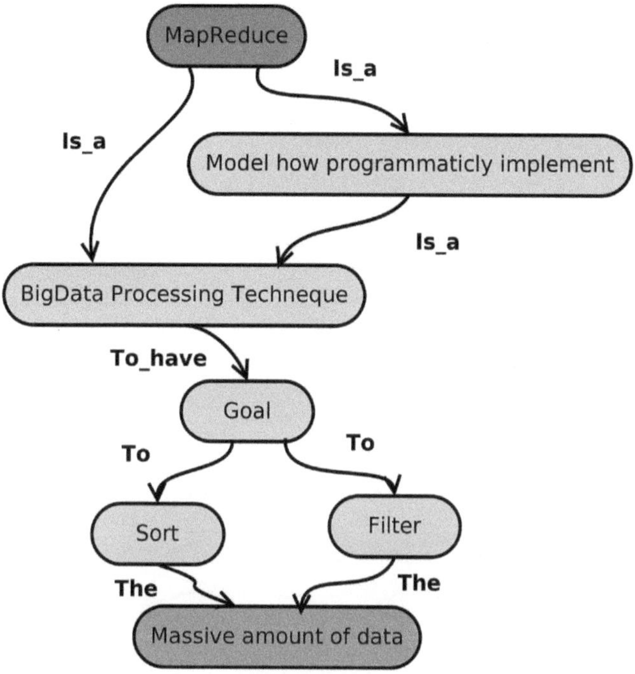

Fig. 4.18 Semantic graph of MapReduce definition.

if the requirement of fast data processing is not critical and a batch mode of working with data is suitable for a business application, as, for example, in ETL (extract, transform, load) systems, then MapReduce provides the required quality of data processing processes.

MapReduce provides fault tolerance and prompt recovery after failures; for example, when a work node that performs a map or reduce operation fails, its work is automatically transferred to another work node if the input data for the operation is available.

Today there are many different commercial and free products that use this distributed computing model: Apache Hadoop, Apache CouchDB, MongoDB, MySpace Qizmt, and other big data frameworks and libraries written in different programming languages (Jose & Abraham, 2020; Mazumdar, Scionti, & Hurson, 2020).

As an example, Fig. 4.19 illustrates the task of collecting data coming from various sensors. It is assumed that the polling time of these sensors can be different and the data enters the database in the form of a sequence of data presented in the form of records.

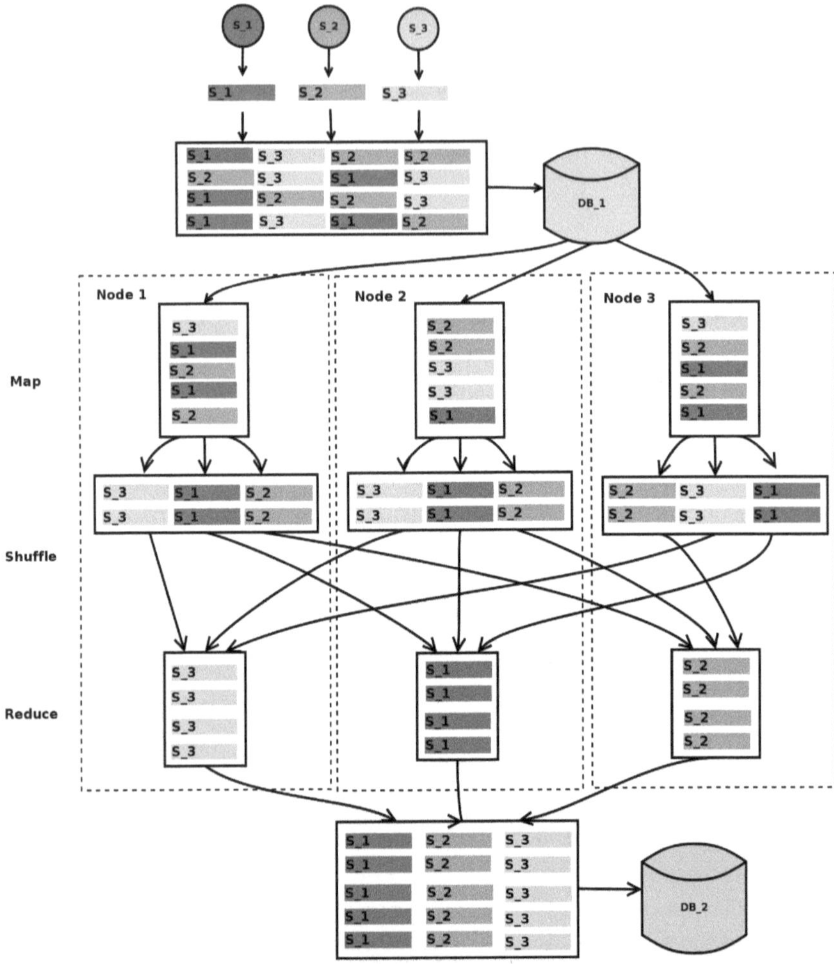

Fig. 4.19 MapReduce processing model.

4.5 Summary

Big data features include:

- data variability associated with changes in transmission speed, formats, structure, semantics, or quality of data arrays;
- the variety of data associated with various formats, logical models, time frames, and semantics of data arrays;
- data transfer rate associated with a high flow rate at which data is created, transmitted, stored, analyzed, or visualized;

- the reliability of the data, determining the completeness and/or accuracy of the data;
- data volatility, reflecting the rate of change of these data over time; and
- the amount of data that affects the computing and storage resources, as well as their management during data processing.

The above features of big data require the use of technologies that provide reliable storage, transmission, and processing of large data arrays. To ensure the processing speed of this data, cluster systems are used in which the process of parallel data processing is implemented.

When preparing data arrays, technologies based on MapReduce are used. The type of data storage model used is of great importance, the choice of which is determined by the features of the tasks being solved.

It should be noted that data processing is implemented on the basis of complex software and hardware systems. Ensuring the safety of processes implemented in them in the era of big data is an important organizational and technical task.

4.6 Definitions

Structured data is data organized on the basis of a given set of rules clearly stated and published in relevant information sources.

Unstructured data is characterized by the absence of any structure other than the structure at the record or file level. An example of unstructured data is text.

Partially structured data is often called semistructured. Examples of partially structured data are records with free text fields in addition to more structured fields.

Metadata is data about data or data items that may include descriptions thereof.

A nonrelational database is a database that does not use a relational model. The terms "NoSQL," "not SQL," and "not only SQL" are used to denote them.

A file is a named record set, considered as a single array of data that can be created, modified, and deleted.

Stream data is data transmitted through the interface from a continuously working data source.

Distributed data processing is data processing in which the execution of operations is distributed between nodes of a computer network.

Parallel data processing is a process in which all calculations are performed in the same time interval on the same computing nodes.

Scatter is a distribution of tasks for data processing among several nodes in a cluster.

Scatter-gather is a method of processing large amounts of data, where the necessary calculations are divided and distributed across several nodes in the cluster, and the overall result is formed by combining the results from each node. An example of data processing using the distribution-assembly method is MapReduce.

Gather is a combination of the results of data processing obtained on several nodes in a cluster.

Vertical scaling (scale-up) is an increase in data processing performance by improving processors, memory, storage, or communications.

Horizontal scaling (scale-out) is the formation of a single logical unit by connecting several hardware and software systems. An example of horizontal scaling is to increase the performance of distributed data processing by adding nodes to the cluster for additional resources.

References

Basford, P. J., Johnston, S. J., Perkins, C. S., Garnock-Jones, T., Tso, F. P., Pezaros, D., … Cox, S. J. (2020). Performance analysis of single board computer clusters. *Future Generation Computer Systems*, *102*, 278–291. https://doi.org/10.1016/j.future.2019.07.040.

Beretta, S., Denti, L., Previtali, M., Ranganathan, S., Gribskov, M., Nakai, K., & Schönbach, C. (2019). *Graph theory and definitions* (pp. 922–927). Oxford: Academic Press. https://doi.org/10.1016/B978-0-12-809633-8.20421-4

Buyya, R., Vecchiola, C., Selvi, S. T., Buyya, R., Vecchiola, C., & Selvi, S. T. (2013). *Principles of parallel and distributed computing* (pp. 29–70). Boston: Morgan Kaufmann. (chapter 2). https://doi.org/10.1016/B978-0-12-411454-8.00002-4.

Buyya, R., Vecchiola, C., Thamarai Selvi, S., Buyya, R., Vecchiola, C., & Selvi, S. T. (2013). *Data-intensive computing: MapReduce programming* (pp. 253–311). Boston: Morgan Kaufmann. (chapter 8). https://doi.org/10.1016/B978-0-12-411454-8.00008-5.

Celko, J., & Celko, J. (2014a). *NoSQL and transaction processing* (pp. 1–14). Boston: Morgan Kaufmann. (chapter 1). https://doi.org/10.1016/B978-0-12-407192-6.00001-7.

Celko, J., & Celko, J. (2014b). *Graph databases* (pp. 27–46). Boston: Morgan Kaufmann. (chapter 3). https://doi.org/10.1016/B978-0-12-407192-6.00003-0.

DB-Engines (n.d.). Retrieved 20 April 2020, from https://db-engines.com/en/.

Dondi, R., Mauri, G., Zoppis, I., Ranganathan, S., Gribskov, M., Nakai, K., & Schönbach, C. (2019). *Graph algorithms* (pp. 940–949). Oxford: Academic Press. https://doi.org/10.1016/B978-0-12-809633-8.20424-X

Guo, X., Du, W., Qi, R., & Qian, F. (2012). Minimum time dynamic optimization using double-layer optimization algorithm. In *Proceedings of the 10th world congress on intelligent control and automation* (pp. 84–88). Beijing. https://doi.org/10.1109/WCICA.2012.6357844.

Hardy, C., & Stobart, S. (2003). *Interacting with the database using SQL* (pp. 187–240). Oxford: Butterworth-Heinemann. (chapter 11). https://doi.org/10.1016/B978-075066076-1/50042-0.

Jose, B., & Abraham, S. (2020). Performance analysis of NoSQL and relational databases with MongoDB and MySQL. In *Vol. 24. International multi-conference on computing, communication, electrical & nanotechnology, I2CN-2K19, 25th & 26th April, 2019* (pp. 2036–2043). https://doi.org/10.1016/j.matpr.2020.03.634.

López, P., & Baydal, E. (2017). On a course on computer cluster configuration and administration. *Keeping up with Technology: Teaching Parallel, Distributed and High-Performance Computing, 105*, 127–137. https://doi.org/10.1016/j.jpdc.2017.01.009.

Mazumdar, S., Scionti, A., & Hurson, A. R. (2020). *Fast execution of RDF queries using Apache Hadoop. Vol. 119* (pp. 1–33). Elsevier. (chapter 1). https://doi.org/10.1016/bs.adcom.2020.03.001.

McKnight, W., & McKnight, W. (2014). *Graph databases: When relationships are the data* (pp. 120–131). Boston: Morgan Kaufmann. (chapter 12). https://doi.org/10.1016/B978-0-12-408056-0.00012-6.

Oshana, R. (2016). Principles of parallel computing. In R. Oshana (Ed.), *Multicore software development techniques* (pp. 1–30). Oxford: Newnes. (chapter 1). https://doi.org/10.1016/B978-0-12-800958-1.00001-2.

Powell, J., Hopkins, M., Powell, J., & Hopkins, M. (2015). Graph databases and how to use them. In *Chandos information professional series* (pp. 197–207). Chandos Publishing. (chapter 22). https://doi.org/10.1016/B978-1-84334-753-8.00022-1.

Scott, M. L. (2009). *Programming language pragmatics* (3rd ed.). Morgan Kaufmann. https://doi.org/10.1016/B978-0-12-374514-9.00010-0.

Sterling, T., Anderson, M., Brodowicz, M., Sterling, T., Anderson, M., & Brodowicz, M. (2018a). *Commodity clusters* (pp. 83–114). Boston: Morgan Kaufmann (chapter 3). https://doi.org/10.1016/B978-0-12-420158-3.00003-4.

Sterling, T., Anderson, M., Brodowicz, M., Sterling, T., Anderson, M., & Brodowicz, M. (2018b). *MapReduce* (pp. 579–589). Boston: Morgan Kaufmann. https://doi.org/10.1016/B978-0-12-420158-3.00019-8

Vaidyanathan, R., Trahan, J. L., Rai, S., Prasad, S. K., Gupta, A., Rosenberg, A. L., … Weems, C. C. (2015). *Introducing parallel and distributed computing concepts in digital logic* (pp. 83–116). Boston: Morgan Kaufmann. (chapter 5). https://doi.org/10.1016/B978-0-12-803899-4.00005-5.

Valeev, S., Kondratyeva, N., & Lutov, A. (2019). Energy consumption optimization of production lines of enterprise for process safety provision. In *2019 international conference on electrotechnical complexes and systems (ICOECS)*. https://doi.org/10.1109/ICOECS46375.2019.8949975.

Venkateswarlu, C., Jujjavarapu, S. E., Venkateswarlu, C., & Jujjavarapu, S. E. (2020). *Stochastic and evolutionary optimization algorithms* (pp. 87–123). Elsevier. (chapter 4). https://doi.org/10.1016/B978-0-12-817392-3.00004-1.

Weilkiens, T., & Oestereich, B. (2007). *UML 2 certification guide*. Germany: Elsevier Inc. https://doi.org/10.1016/B978-0-12-373585-0.X5000-4.

Witten, I. H., Frank, E., Hall, M. A., & Pal, C. J. (2017). *Algorithms: The basic methods* (4th ed., pp. 91–160). Morgan Kaufmann. (chapter 4). https://doi.org/10.1016/B978-0-12-804291-5.00004-0.

CHAPTER 5

Simulation technologies for process safety

5.1 Simulation of process safety

Modeling is one of the main methods for the study of complex organizational and technical systems in various fields of human activity, including the rapidly growing information sphere. The simulation results are used for management (decision making) in both existing and designed systems. At the present stage of development of information technology, computer mathematical modeling has become the most widespread. The use of computer technology allows the effective combination of simulation and analytical models (Manson, 2020; Zhang, Zeigler, & Laili, 2019).

In a broad sense, modeling can be defined as a method of cognition, in which the studied original object is replaced by a model object, and the model reflects those characteristics of the object that are significant from the point of view of a particular cognitive process. Thus, modeling can be defined as the representation of an object by a model to obtain information about this object by conducting experiments with its model.

By simulation we understood a step-by-step reproduction of the process of functioning of a system in time. It is advisable to use simulation models if the modeling object has a complex heterogeneous structure, and the construction of a complete analytical model for it is impossible (Saeid, Poe, & Mak, 2019; Salimi & Salimi, 2018; Zhang, Wu, Durand, Albalawib, & Christofides, 2018).

The class of such complex objects that are difficult to formalize can include the majority of modern organizational and technical systems, including process safety. Moreover, any experiment on the model in order to study the properties of the simulated system can be of significant importance only with special preparation, processing, and generalization of the results (Fig. 5.1).

In general, in order to build models and conduct model experiments successfully, knowledge of the theory of systems, theory of statistics, standard mathematical schemes, as well as the basics of experimental design is necessary.

Process Safety and Big Data
https://doi.org/10.1016/B978-0-12-822066-5.00006-6

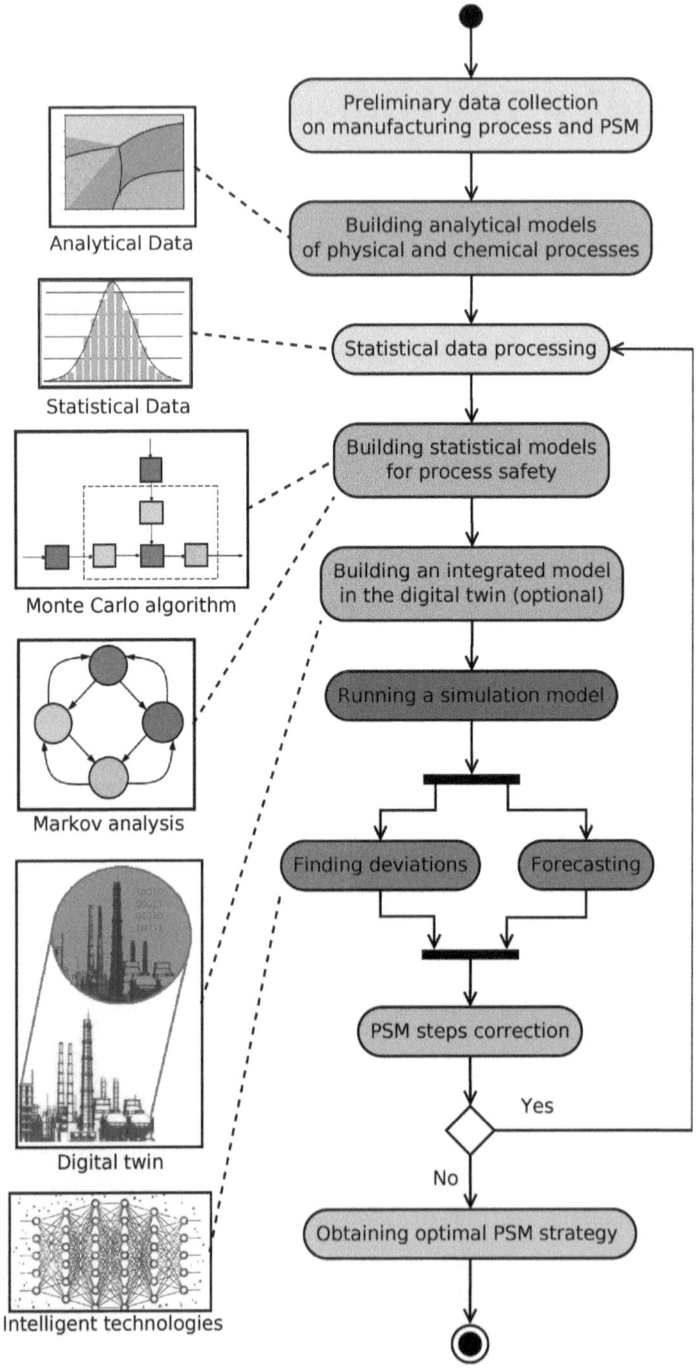

Fig. 5.1 Applying simulation technologies for optimization of process safety management (PSM).

A model can be developed using a deterministic approach based on mathematical equations. When a system is complex and functions under conditions of significant uncertainty, it is not always possible to describe it using classical physical or chemical laws. In this case, stochastic or probabilistic models can be used, for instance Markov models.

IEC 31010:2019 recommends using modeling including software models to analyze collected information for the process of risk assessment. A model represents a real system or process in a simplified form. It aims to transform a complex system into simplified objects that are easier to analyze. It can be used to understand the meaning of data and to simulate what can happen in reality under various conditions. Software can be used effectively to represent, organize, and analyze data.

5.1.1 Accuracy of process parameters simulation

Modern process industries operate under the influence of a large number of random factors. Examples of such factors include equipment uptime, repair time, moments of requests for information processing, etc. Due to the great complexity of objects of this class, models can also have a complex hierarchical structure, often built into the framework of digital twins. When implementing such complex models, taking into account uncertainties, the accuracy of the results is of great importance (Galdi & Tagliaferri, 2019; Wang, Wassan, & Zheng, 2018). Accuracy is one metric for evaluating classification models. Informally, accuracy is the fraction of predictions the model gets right.

If we classify the errors in the implementation of the "ideal" computer model from the point of view of the causes of their occurrence, we can distinguish the following main groups (Carpenter, 2020; Hartmann, 2012):

- modeling errors arising due to inaccurate input data (or ignorance of their nature);
- modeling errors resulting from the simplification of the original simulation model;
- errors in the calculation of state variables and model output parameters due to discrete implementation of the simulation model; and
- modeling errors due to the limited amount of statistical data or a limited number of random model tests on a computer.

We briefly consider individual groups of errors below.

Modeling errors arising due to inaccurate input data have the following particularities. In the general case, by their nature, the input factors of the simulation model can be divided into controlled variables (chosen by the researcher), deterministic, random, and indefinite. Taken into account in the model, even a very large number of determinate factors does not lead to significant computational difficulties and errors. The inclusion of random factors in the model by two to three orders of magnitude increases the amount of computation. An increase in the number of variables and uncertain factors in optimization simulation models also significantly increases the amount of computation for finding optimal solutions. In some cases, their large dimension does not allow us to find the optimal (rational) solution for the admissible time.

To reduce the amount of computation, the researcher, as a rule, seeks to consider some random and uncertain factors as determinate, thereby introducing errors into the simulation results. In addition, a priori ignorance of information about the object (or its inaccuracy) leads to the fact that the numerical specification of the initial data of the model (initial data in the form of constants) will be made with errors. Errors of this type are usually estimated in advance. The researcher must know the price of a particular replacement. To study the influence of these errors on the accuracy of modeling results, as a rule, special methods of sensitivity theory are used.

Modeling errors resulting from the simplification of the original simulation model arise for the following reasons. The initial simulation model, as a rule, is simplified to obtain even an approximate but analytical solution that allows one to determine quickly both the region of optimal parameters and the influence of various model factors on this region. Such procedures are carried out, for example, by replacing nonlinear dependencies with linear ones, polynomials of high degrees with polynomials of low degrees, nonsmooth functions with smooth ones, etc. The error value of such transformations must also be calculated in advance.

Errors in the calculation of state variables and model output parameters due to the discrete implementation of the simulation model give the following types of errors:

- errors in the calculation of state variables and model output parameters associated with the implementation of simulation mechanisms for the processes under study (selection of principles for constructing modeling algorithms);
- rounding errors of intermediate results;
- errors associated with replacing an infinite computational process with a finite one, for example, the derivative is replaced by a finite difference, the integral is replaced by a sum, etc. (these are methodological errors of ordinary numerical methods); and

- errors associated with the replacement of continuous values by discrete ones in the numerical study of processes; the error depends on the sampling step.

When developing simulation models, it is necessary to choose such discrete implementation methods that, based on the available information, allow us to state that the modeling errors will not exceed the specified values.

Errors due to the limited volume of statistical data are characteristic of simulation models that include random factors in the input data. The researcher always deals with a limited statistical sample, in contrast to the general set of statistical data. In this regard, the form and characteristics of distribution laws will vary; the magnitude of this discrepancy (error) depends on the size of the statistical sample.

For simulation modeling, the resulting error depends both on the volume of experimental data on the values of the studied random variables and on the number of realizations—model runs for different values of random variables. A measure of their quantitative expression is the value of the confidence interval of certain characteristics of the experiment (the value of the confidence interval is calculated and set at the planning stage of simulation experiments). These errors, as a rule, are controlled by the researcher in the sense that in the process of planning the experiment, changing them within reasonable limits of the confidence interval, you can get an allowable error of the simulation results.

5.1.2 Simulation algorithms

Various algorithms based on probability theory and mathematical statistics can be used to implement the developed models (computer simulation) of the process industry in the form of digital twins using big data. The sequence of implementation of the model, in general, will be as follows:

- collection of raw field data about the research objects (enterprise);
- collection of data on similar facilities (fleet of facilities);
- statistical processing of collected historical data or a temporary cut of the state of a group of similar objects; and
- modeling of initial data in the form of obtained statistics and processes of the system's functioning using a computer in order to identify the features of the system in various conditions, optimizing the functioning of the system and minimizing the risk to process safety.

Monte Carlo analysis is widely used to model input random influences (source data), and Markov models are used to study the processes of the functioning of the system.

These two methods were already mentioned in the second chapter when considering the IEC 31010:2109 standard as recommended techniques for use in risk analysis, including process safety risk. Next, we consider these modeling methods in more detail.

The Monte Carlo method (method of statistical experiments) is widely used to model random input variables at the stage of research and design of systems. This method is based on the use of random numbers, that is, possible values of a random variable with a given probability distribution (ISO/IEC Guide 98-3:2008/Suppl 1: Uncertainty of measurement—Part 3: Guide to the expression of uncertainty in measurement (GUM 1995)—Propagation of distributions using a Monte Carlo method, 2008; Kroese, Brereton, Taimre, & Botev, 2014). Statistical modeling is a method of obtaining, using a computer, statistical data on the processes occurring in a simulated system. To evaluate the characteristics of a simulated system of interest, taking into account the effects of environmental factors, statistical data are processed and classified using mathematical statistics methods.

The essence of the statistical modeling method is to build for the functioning of the system under study S a certain modeling algorithm that simulates the behavior and interaction of system elements taking into account random input influences and environmental influences E and the implementation of this algorithm using software and hardware.

There are two areas of application of the statistical modeling method: (1) for the study of stochastic systems; (2) for solving deterministic problems. The main idea that is used to solve deterministic problems by the method of statistical modeling is to replace the determinate problem with an equivalent circuit of some stochastic system (Gonzaga, Arinelli, de Medeiros, & Araújo, 2019; Hastings, 1970). The output characteristics of the latter coincide with the solution to the deterministic problem. Obviously, with such a replacement, instead of the exact solution to the problem, an approximate solution is obtained, and the error decreases with increases in number of tests (implementations of the modeling algorithm) N.

As a result of statistical modeling of the system S, a series of particular values of the sought quantities or functions is obtained, the statistical processing of which provides information about the behavior of a real object or process at arbitrary points in time. If the number of realizations N is sufficiently large, then the obtained results of system modeling acquire statistical stability and can be taken with sufficient accuracy as estimates of the desired characteristics of the process of functioning of system S.

The theoretical basis of the method of statistical computer-based simulation of systems is the limit theorems of probability theory (Chung, 2014).

Statistical computer simulation requires the generation of random values. This can be done using sensors (generators) of random numbers. Consider a simple example of constructing and implementing a model using the Monte Carlo method:

Example

It is necessary by statistical modeling to find estimates of the output characteristics of a certain stochastic system S_R, the functioning of which is described by the following relationships: $x = 1 - e^{-1}$ is input action, $v = 1 - e^{-j}$ is the environmental impact, where 1 and j are random variables, for which their distribution functions are known. The purpose of the simulation is to estimate the expected value $E[Y]$ of the variable Y. The dependence of the latter on the input action x and the influence of the external environment v has the form $Y = \sqrt{x^2 + v^2}$.

As an estimate of the expected value $E[Y]$, as follows from the theorems of probability theory, the arithmetic mean calculated by the formula $\overline{Y} = {}^1/_N \sum_{i=1}^{N} Y_i$ where Y_i is a random value of y; N is the number of realizations necessary for the statistical stability of the results. The block diagram of the S_R system is shown in Fig. 5.2.

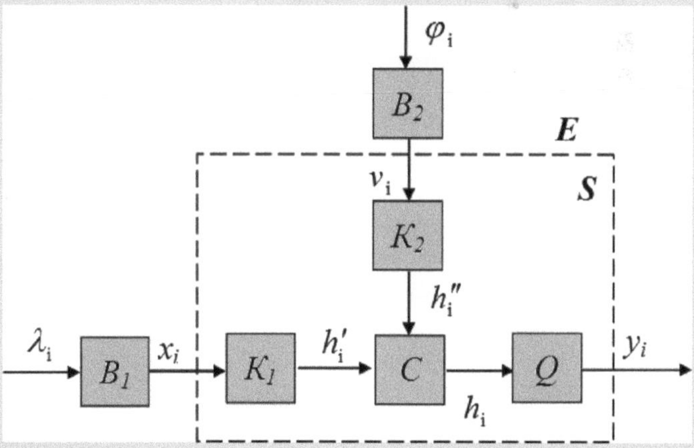

Fig. 5.2 System SR block diagram.

Here, the elements perform the following functions:

- calculation $B_1 : x_i = 1 - e^{-\lambda_i}$ and $B_2 : v_i = 1 - e^{-\varphi_i}$;
- squaring $K_1 : h_i' = (1 - e^{-\lambda_i})^2$ and $K_2 : h_i'' = (1 - e^{-\varphi_i})^2$;

- summation $C: h_i = (1 - e^{-\lambda_i})^2 + (1 - e^{-\varphi_i})^2$; and
- square root extraction $Q: Y_i = \sqrt{(1 - e^{-\lambda_i})^2 + (1 - e^{-\varphi_i})^2}$.

The scheme of the algorithm that implements the statistical modeling method for estimating $E[y]$ of the $\boldsymbol{S}_R$ system is shown in Fig. 5.3.

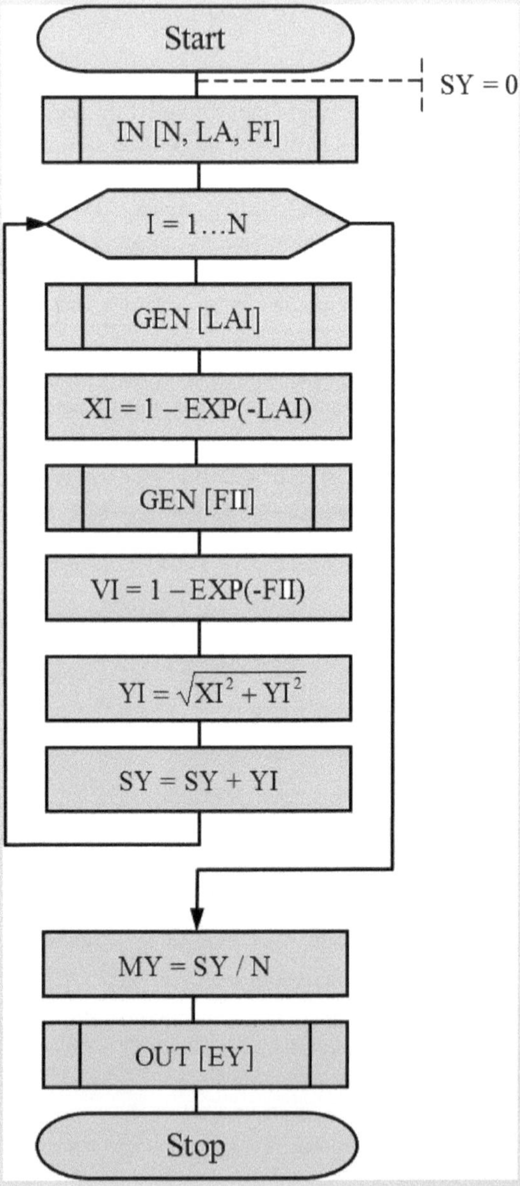

Fig. 5.3 The scheme of the modeling algorithm for the system SR.

Here LA and FI are the distribution functions of the random variables λ and φ; N is a given number of model implementations; $I \equiv i$ is the number of the current implementation; $LAI \equiv \lambda_i$; $FII \equiv \varphi_i$; EXP$=e$; $MY = M[y]$; $SY = \sum_{i=1}^{N} Y_i$ is the summing cell; and IN [...], GEN [...], and OUT [...] are procedures for inputting initial data, generating pseudorandom sequences, and issuing simulation results, respectively.

Thus, this model makes it possible to obtain, by statistical modeling on a computer, a statistical estimate of the expected value of the output characteristic $E[y]$ of the considered stochastic system $\boldsymbol{S}_R$. The accuracy and reliability of the simulation results will mainly be determined by the number of implementations N.

In the considered example, it is not necessary to memorize the entire set of generated random variables used in statistical modeling of the $\boldsymbol{S}_R$ system, and this is typical for this class of simulation models as a whole.

5.1.2.1 Simulation of random events

The simplest random objects in statistical modeling of systems are random events. Consider the features of their modeling. Probability basics were discussed at the beginning of the second chapter, so please go back to refresh knowledge on probabilities, events, and experiments (tests).

Let there be random numbers x_i, i.e., possible values of a random variable ξ uniformly distributed in the interval (0,1). It is necessary to realize a random event A occurring with a given probability p. We define A as an event consisting in the fact that the selected value of x of a random variable ξ satisfies the inequality:

$$x_i \leq p. \tag{5.1}$$

Then the probability of event A will be $P(A) = \int_0^p dx = p$. The opposite event $\overline{A}$ is that $x_i > p$. Then $P(\overline{A}) = 1 - p$. The modeling procedure in this case consists in choosing the values of x_i. and comparing them with p. Moreover, if condition (5.1) is satisfied, the outcome of the test is event A.

In the same way, we can consider a group of events. Let $A_1, A_2, \ldots, A_s$ be the complete group of events occurring with probabilities $p_1, p_2, \ldots, p_s$, respectively. We define A_m as an event consisting in the fact that the chosen value x_i of the random variable ξ satisfies the inequality:

$$l_{m-1} < x_i \leq l_m, \tag{5.2}$$

where $l_m = \sum_{i=1}^{m} p_i$.

Then $P(A_m) = \int_{l_{m-1}}^{l_m} dx = p_m$.

The test simulation procedure in this case consists in sequentially comparing random numbers x_i, with l_m values. The outcome of the test is the event A_m, if condition (5.2) is satisfied.

This procedure is called *determining the outcome of the draw test* in accordance with the probabilities $p_1, p_2, \ldots, p_s$. These modeling procedures were considered under the assumption that random numbers x_i with a uniform distribution in the interval (0,1) are used for testing.

When modeling systems, it is often necessary to carry out such tests, in which the desired result is a complex event, depending on two (or more) simple events. Let, for example, independent events A and B have the probability of occurrence of p_A and p_B. Possible outcomes of compatible tests in this case would be events $AB, \overline{A}B, A\overline{B}, \overline{AB}$ with $p_A p_B$, $(1-p_A)p_B$, $p_A(1-p_B)$, and $(1-p_A)(1-p_B)$.

To simulate compatible tests, two versions of the procedure can be used: (1) sequential verification of condition (5.1); (2) determination of one of the outcomes $AB, \overline{A}B, A\overline{B}, \overline{AB}$, by lot with corresponding probabilities, i.e., analogy (5.2). The first option requires two numbers x_i and comparisons to verify condition (5.1). The second option can be performed with one number x_i, but comparisons may require more. From the point of view of the convenience of constructing a modeling algorithm and saving the number of operations and computer memory, the first option is preferable.

We now consider the case when events A and B are dependent and occur with probabilities p_A and p_B. We denote by $P(B/A)$ the conditional probability of the occurrence of event B, provided that event A has occurred. Moreover, we consider that the conditional probability $P(B/A)$ is given.

Consider one of the options for building a model. The next number x_m is extracted from the sequence of random numbers $\{x_i\}$ and the validity of the inequality $x_m < p_A$ is checked. If this inequality is true, then event A has occurred. For the test associated with event B, the probability $P(B/A)$ is used. The next number x_{m+1} is taken from the set of numbers $\{x_i\}$ and the condition $x_{m+1} \leq P(B/A)$ is checked. Depending on whether or not this inequality holds, the outcome of the test is AB; or, if the inequality $x_m < p_A$ is not satisfied, then the event $\overline{A}$ has occurred. Therefore, for the test associated with event B, it is necessary to determine the probability $P(B|\overline{A}) = [P(B) - P(A)P(B|A)]/(1-P(A)]$.

From the set $\{x_i\}$ we choose the number x_{m+1} and verify the inequality $x_{m+1} \leq P(B/\overline{A})$. Depending on whether it is performed or not, we obtain the outcome of the test $\overline{A}B$ or $\overline{A}\,\overline{B}$. The logic diagram of the algorithm for implementing this version of the model is shown in Fig. 5.4.

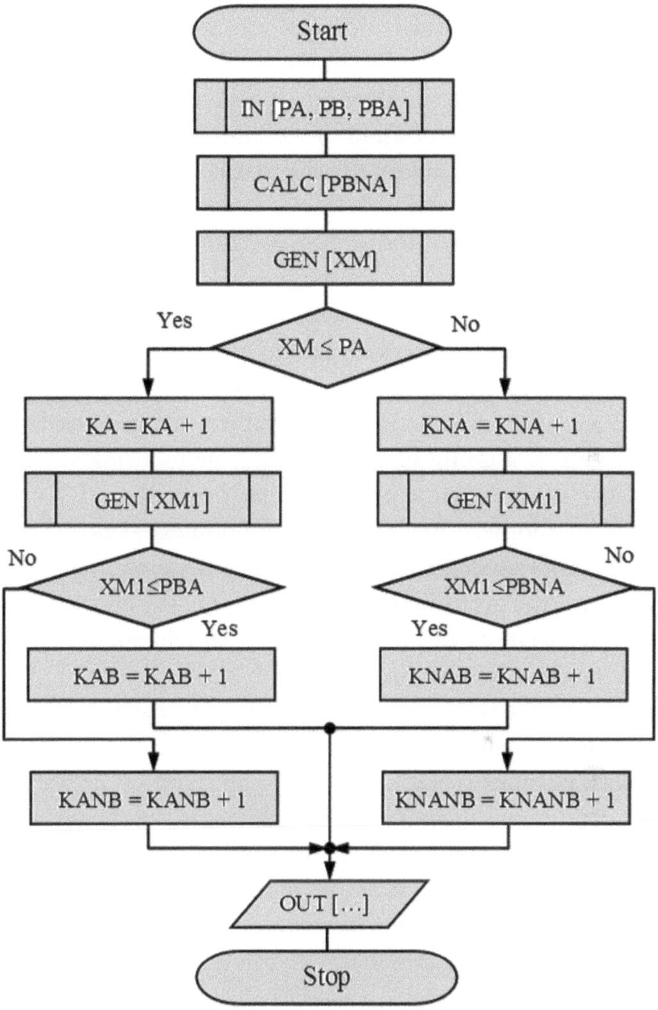

Fig. 5.4 Scheme of the modeling algorithm for dependent events.

Here IN [...] is the input data input procedure; CALC [...] is the procedure for calculation the conditional probability of the occurrence of event B, provided that event A has occurred $P(B|A)$; GEN [...] is a generator of uniformly distributed random numbers; $XM^{\circ} \equiv x_m$; $XM1^{\circ} \equiv x_{m+1}$; $PA^{\circ} \equiv p_A$; $PB^{\circ} \equiv p_{\mathrm{B}}$; $PBA^{\circ} \equiv P(B|A)$; $PBNA \equiv P\left(B|\overline{A}\right)$; KA, KNA, KAB, $KNAB$, $KANB$, $KNANB$ are the number of events A, $\overline{A}$, AB, $\overline{A}B$, $A\overline{B}$, $\overline{AB}$, respectively; and OUT [...] is the procedure for issuing simulation results.

For the formation of possible values of *random variables* with a given distribution law of the source data, basic sequences of random numbers $\{x_i\}$ are used, having a uniform distribution in the interval (0, 1).

In other words, random numbers x_i as possible values of a random variable ξ having a uniform distribution in the interval (0, 1) can be converted into possible values y_j of a random variable η, the distribution law of which is given.

5.1.2.2 Markov models

Let there be some physical system S, which, over time, passes from one state to another in a previously unknown, random manner. In other words, a random process takes place in the system.

As a system, objects of various physical natures can be considered: an oil platform, a refinery, a group of such plants, a petrochemical enterprise, a processing industry, etc. Most processes that occur in real systems have, to one degree or another, the features of randomness—uncertainties.

A random process occurring in the system is called *Markov* if for any moment in time t_0 the probabilistic characteristics of the process in the future depend only on its state at the given moment t_0 and do not depend on when and how the system came to this state.

Let at the moment t_0 the system be in a certain state S_0. At this point in time, the state of the system S_0 and the entire history of the process for $t < t_0$ are known. Is it possible to predict the future behavior of the system ($t > t_0$) under these conditions? With absolute accuracy, it is impossible, since the process is by definition random, therefore unpredictable. But some probabilistic characteristics of the process in the future can still be found. For example, the probability that after some time t the system S will be in state S_1 or save state S_0, etc.

Of great importance are the so-called *Markov random processes with discrete states and continuous time* (Hajek, 2015; IEC 61165:2006 Application of Markov techniques, 2006; Karlin & Taylor, 1975; Meyn & Tweedie, 2009). A process is called a process with discrete states if its possible states S_1, S_2, S_3, ... can be enumerated (renumbered) in advance, and the transition of the system from state to state occurs "in a jump," almost instantly. A process is called a process with continuous time if the moments of possible transitions from state to state are not fixed in advance, but are uncertain, random, if the transition can occur, in principle, at any time.

To construct a mathematical model of the Markov process, the concept of *a flow of events* is also required.

A flow of events is a sequence of homogeneous events that follow one after another at some random times. For example: a stream of alarms in a situation center, a stream of equipment failures (malfunctions), a stream of trains arriving at a sorting station, etc.

The intensity of the event flow λ is the average number of events per unit time. The flow of events is called *stationary* if its probabilistic characteristics are time-independent. In particular, the flow intensity λ must be constant. The flow of event is called *the flow without consequences* if for two disjoint time intervals τ_1 and τ_2 the number of events falling on one of them does not depend on how many events fall on the other. This means that the events that make up the flow appear at different times independently of each other. The flow of events is called *ordinary* if events in it appear one at a time and not in groups at once. For example, the flow of trains approaching the station is ordinary, but the flow of wagons making up those trains is not. The flow of events is called the *simplest* if it has three properties simultaneously: stationary, ordinary, and has no consequences. For the simplest flow with intensity λ, the interval T between adjacent events has the so-called exponential distribution with density $f(t)$ (Fig. 5.5):

$$f(t) = \lambda e^{-\lambda t}, \quad t > 0. \tag{5.3}$$

The variable λ in formula (5.3) is called the parameter of the exponential law.

Considering Markov processes with discrete states and continuous time, it is convenient to assume that all transitions of the system S from state to state occur under the influence of some event flows (call flow, failure flow,

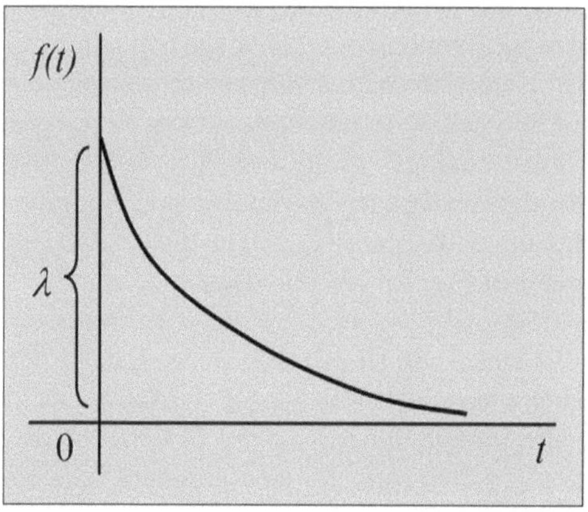

Fig. 5.5 Exponential distribution density.

recovery flow, etc.). If all the flows of events that transfer system *S* from state to state are the simplest, then the process taking place in the system will be Markov. This is natural, since the simplest flow has no aftereffect, in that the "future" does not depend on the "past" (Gagniuc, 2017; Gamerman & Lopes, 2006).

If the system *S* is in some state S_i, from which there is a direct transition to another state S_j, then we will imagine it as if the system, while it is in state S_i, is affected by the simplest flow of events, moving it in the direction of the arrow $S_i \rightarrow S_j$. As soon as the first event of this flow appears, the system transitions from S_i to S_j. For clarity, it is very convenient on the state graph for each arrow to put down the intensity of the stream of events that translates the system along this arrow (λ_{ij}).

Example

The technical device *S* consists of two units, each of which at a random moment in time can fail, after which the repair of the unit, also continuing in advance of an unknown, random time, immediately begins. The following system states are possible: S_0—both units are operational; S_1—the first unit is being repaired, the second unit is in good condition; S_2—the second unit is being repaired, the first unit is working; S_3—both units are being repaired. We construct a signed state graph for this example in Fig. 5.6.

We will calculate the intensities of the event flows that transfer the system from state to state assuming that the average repair time of a unit does not depend on whether one unit is being repaired or both at once.

This will be the case if a separate specialist is busy repairing each unit. Find all the intensities of the flows of events that transfer the system from state to state. Let the system be in state S_0. What stream of events translates it to state S_1? Obviously, it will be the failure flow of the first node.

Its intensity λ_1 is equal to unity divided by the average uptime of the first unit. What event flow takes the system back from S_1 to S_0? Obviously, it is the stream of "repair endings" of the first unit. Its intensity μ_1 is equal to unity divided by the average repair time of the first unit. Similarly, the intensities of the flows of events that translate the system along all the arrows in the graph of Fig. 5.6 are calculated.

Having at its disposal a signed state graph of the system, it is easy to construct a mathematical model of this process. Indeed, let a system *S* be considered, having *n* possible states $S_1, S_2, \ldots, S_n$. We call the probability of the *i*th state the probability $p_i(t)$ of the fact that at time *t* the system will be in state S_i. Obviously, for any moment, the sum of all state probabilities is equal to unity: $\sum_{i=1}^{n} p(t) = 1$.

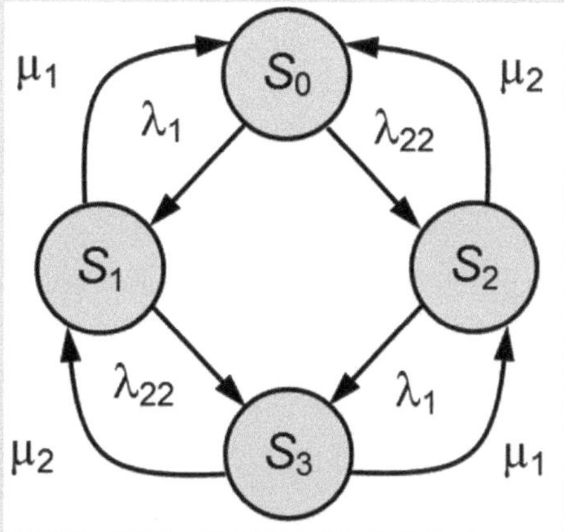

Fig. 5.6 The signed state graph for the system *S*.

Having at our disposal a labeled state graph, we can find all the probabilities of the states $p_i(t)$ as a function of time. To do this, the so-called Kolmogorov equations, which are special types of differential equations in which the state probabilities are unknown functions, are compiled and solved.

We formulate the general rule for compiling Kolmogorov equations. On the left side of each of them is the derivative of the probability of some (*i*th) state. On the right side is the sum of the products of the probabilities of all states from which the arrows go to a given state, by the intensity of the corresponding event flows, minus the total intensity of all flows that remove the system from this state, multiplied by the probability of this (*i*th) state.

Using this rule, we write the Kolmogorov equations for the system *S*, the signed state graph of which is shown in Fig. 5.6:

$$\left.\begin{aligned} \frac{dp_0}{dt} &= \mu_1 p_1 + \mu_2 p_2 = (\lambda_1 + \lambda_2) p_0, \\ \frac{dp_1}{dt} &= \lambda_1 p_0 + \mu_2 p_3 - (\lambda_2 + \mu_2) p_1, \\ \frac{dp_2}{dt} &= \lambda_2 p_0 + \mu_1 p_3 - (\lambda_1 + \mu_2) p_2, \\ \frac{dp_3}{dt} &= \lambda_2 p_0 + \lambda_1 p_2 - (\mu_1 - \mu_2) p_3. \end{aligned}\right\} \tag{5.4}$$

To solve the Kolmogorov equations and find the probabilities of states, first of all, you need to set the initial conditions. If we know exactly the initial state of the system S_i, then at the initial moment $(t=0)$ $p_i(0)=1$, and all other initial probabilities are equal to zero. So, for example, it is natural to solve equations (5.4) under the initial conditions $p_0(0)=1$, $p_1(0)=p_2(0)=p_3(0)=0$ (at the initial moment both nodes are operational).

The obtained linear differential equations with constant coefficients can be solved analytically, but this is convenient only when the number of equations does not exceed two (sometimes three). If there are more equations, then they are usually solved numerically, i.e., manually or on a computer.

For $t \to \infty$ in the system S, a limit stationary regime is established in which the system randomly changes its states, but their probabilities are no longer time-dependent. The final probability of the state of S_i can be interpreted as the average relative residence time of the system in this state. For example, if the system S has three states S_1, S_2, S_3, and their final probabilities are 0.2, 0.3, and 0.5, this means that in the limit stationary state the system spends on average two tenths of the time in state S_1, three tenths—in state S_2 and half the time—in state S_3.

Since the final probabilities $p_1, p_2, \ldots$ are constant, their derivatives are equal to zero. Therefore, in order to find the final probabilities, it is necessary to put all the left parts in the Kolmogorov equations equal to zero and solve the resulting system of linear algebraic equations, not differential ones. We do not need to write Kolmogorov equations, but write a system of linear algebraic equations directly from the state graph. If we transfer the negative term of each equation from the right side to the left, we immediately get a system of equations where the final probability of a given state pi stands on the left, multiplied by the total intensity of all flows leading from this state, and on the right is the sum of the products of the intensities of all flows included in i^{th} state, on the probability of those states from which these flows emanate.

Using this rule, we write linear algebraic equations for the final probabilities of the states of a system whose state graph is shown in Fig. 5.6:

$$\left.\begin{aligned}(\lambda_1+\lambda_2)p_0&=\mu_1 p_1+\mu_2 p_2,\\(\lambda_2+\mu_2)p_1&=\lambda_1 p_0+\mu_2 p_3,\\(\lambda_1+\mu_2)p_2&=\lambda_2 p_0+\mu_1 p_3,\\(\mu_1+\mu_2)p_3&=\lambda_2 p_0+\lambda_1 p_2.\end{aligned}\right\} \tag{5.5}$$

It would seem that this system of four equations with four unknowns p_0, p_1, p_2, p_3 can be completely solved. However, equations (5.5) are homogeneous (do not have a free term) and, therefore, the unknowns determine only up to an arbitrary factor. Fortunately, we can use the so-called normalization condition:

$$p_0+p_1+p_2+p_3=1 \tag{5.6}$$

and use it to solve the system. Moreover, one (any) of the equations can be discarded (it follows as a consequence of the rest). Let's set the numerical values of the intensities $\lambda_1=1$, $\lambda_2=2$, $\mu_1=2$, $\mu_2=3$ and solve the system (5.6). We sacrifice the fourth equation by adding the normalizing condition (5.5) instead. Then the equations will take the form:

$$\left.\begin{aligned} 3p_0 &= 2p_1 + 3p_2, \\ 4p_1 &= p_0 + 2p_3, \\ 4p_2 &= 2p_0 + 2p_3, \\ p_0 + p_1 + p_2 + p_3 &= 1. \end{aligned}\right\} \tag{5.7}$$

Solving it, we shall obtain the following: $p_0 = 6/15 = 0.40$; $p_1 = 3/15 = 0.20$; $p_2 = 4/15 \approx 0.27$; $p_3 = 2/15 \approx 0.13$, i.e., in the limit or stationary mode, the system S will spend on average 40% of the time in state S_0 (both nodes are operational), 20% in state S_1 (the first node is being repaired, the second is working), 27% in state S_2 (the second node is being repaired, the first is working), and 13% in a state of S_3 in complete disrepair (both nodes are being repaired). Knowing these final probabilities can help evaluate the average system performance and the load of repair bodies. Suppose that a system S in state S_0 (fully functional) yields income 8 (conventional units) per unit time, in state S_1—income 3, in state S_2—income 5, in state S_3—does not generate income at all. Then in the limit stationary mode, the average income per unit time will be $W = 0.4 * 8 + 0.20 * 3 + 0.27 * 5 = 5.15$.

Now we estimate the load of repair staff (workers) engaged in the repair of units 1 and 2. Unit 1 is being repaired for a part of the time equal to $p_1 + p_3 = 0.20 + 0.13 = 0.33$. Unit 2 is repaired over a part of the time $p_2 + p_3 = 0.40$. Here a question may already arise about optimizing the solution. Suppose that we can reduce the average repair time of one or the other unit (maybe one or the other), but it will cost us some amount. Will the increase in income associated with faster repairs pay for the increased repair costs?

5.2 Digital twins

The use of digital twins is a modern and promising area of sharing modeling and big data technologies within the framework of the "Industry 4.0" concept (Erboz, 2017; Gronau, Grum, & Bender, 2016). Industry 4.0 involves the massive introduction of cyber-physical systems (CPS) in various areas of human activity, including manufacturing and process industries.

A cyber-physical system is an information technology system that implies the integration of computing resources into physical objects of any nature, including both biological and artificially created objects. In cyber-physical systems, computing components are distributed throughout the physical system, which at the same time is its carrier and are synergistically linked to the elements of this system (Lee & Seshia, 2011; Sanfelice, 2016).

5.2.1 Digital twins for process safety

A *digital twin* is a digital (virtual) copy of a physical (real) system or process created to optimize system performance. If we talk about the process industry and process safety, the purpose of the digital twin in this area is the early detection of deviations of production process parameters from the set, prediction and prevention of critical situations, support for decision-making on process safety management (PSM), production of better products, risk reduction, and safety of people and the environment (El Saddik, 2018; Tao et al., 2018).

On the one hand, a digital twin can be an integrated model of an already created production facility. Such a model should contain information about all failures in the operation of the enterprise and be regularly updated during operation. Information for updating the model comes from sensors installed on a physical object. This allows you to simulate the behavior of an object in real time (Fig. 5.7).

On the other hand, a digital twin can be defined as a constantly changing digital profile containing historical and most relevant data about a physical

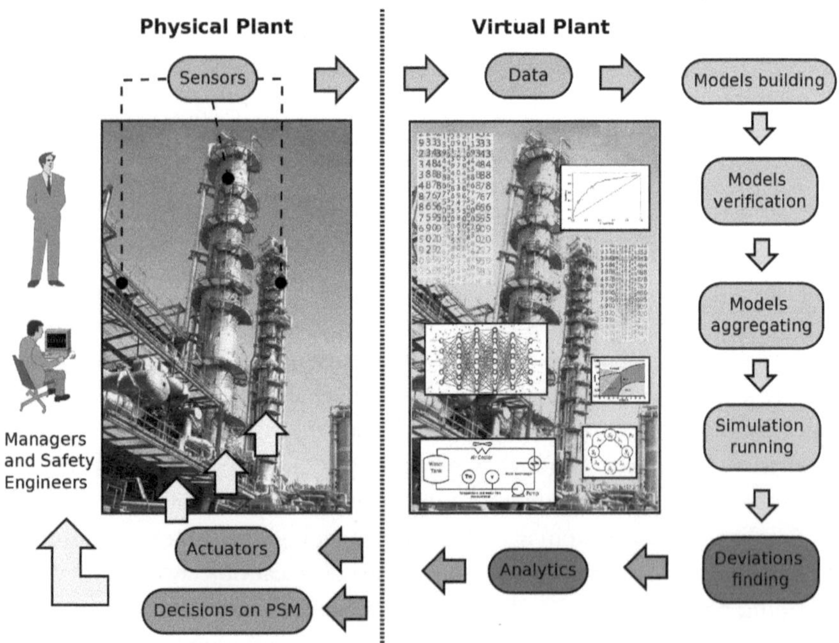

Fig. 5.7 Digital twin for process safety management (PSM).

object or process, which allows the efficiency of production facilities to be increased, including ensuring process safety. It should also be noted that only significant progress in the development of digital technologies, which allowed to increase computing power and reduce the cost of their use, made it possible to combine information technology with the operational processes of enterprises to create digital twins. The development of a digital twin is based on the use of a huge amount of accumulated data obtained during measurements of numerous indicators of a real (physical) object. Analysis of the accumulated data makes it possible to obtain accurate information about the current efficiency of the production system, as well as to support decision-making on the need to make changes to both the production process and process safety management (PSM).

Most often, digital twins are created to simulate objects directly related to industrial production (Yi, Binil, Paul, & Yuan-Shin, 2017). For example, to simulate a tank with a valve and a pump, filling sensors as well as sensors on actuators are used as a data source, which allow simulating processes using SysML, AML, SCADA, and ANFIS. In addition, the experimental data obtained in the course of measuring the deformation of parts and components of the pyrolysis unit allow us to simulate its state using realistic SIMULIA simulation applications. In addition, it is possible to simulate organizational processes in an enterprise to develop management decisions in critical situations as part of PSM.

The digital twin application constantly analyzes incoming data streams. Over a period of time, the analysis may reveal unacceptable trends in the actual productivity of the production process in a particular dimension compared to the ideal range of acceptable characteristics. Such a comparative understanding can provoke research and a potential change in some aspects of the production process in the physical world.

Sensors distributed throughout the production process generate signals that allow the twin to collect work and environmental data related to the physical process in the real world.

Thousands of sensors carry out continuous nontrivial measurements that are transmitted to a digital platform, which, in turn, performs an analysis in almost real time to optimize the production process.

Actual operational and environmental data from the sensors are combined, and combined with data from the plant, such as bills of materials (BOM), plant systems, and project specifications. The data may also contain other elements, such as technical drawings, connections to external data sources, and logs of critical situations and incidents.

Sensors transmit data to the digital world using integration technology (which includes boundaries, communication interfaces, and security) between the physical world and the digital world, and vice versa.

Analytical methods are used to analyze data using algorithmic modeling and visualization procedures that are used by the digital twin to obtain information. If any action is justified in the real world, the digital twin performs the action using actuators that are subject to human intervention, or managerial decisions are made that trigger the physical process.

Summarizing the above, we can say that for the use of digital twins in process safety, the digitization and integration of two key areas—namely risk management and operational management—is necessary. First, you need to digitize the source data and recommendations of security engineering experts (HAZOP, LOPA, PHA, SRS, SIL/SIF calculations, etc.), followed by entering into a digital database that ensures consistency, speed of search, and ease of use of critically important process safety information.

Then it is necessary to compare the digital "design" data with the actual "operational" data and, based on the simulation, find out the impact on the risk of possible system degradation in real time. To do this, it is necessary to minimize manual data processing and regularly supplement the generated reports with real-time information.

After that, big data analytics are used to extract knowledge from heterogeneous process safety data, comparing the actual "operational" conditions (for example, the speed of requests to the sensors, test intervals, bypass time) with the "design" data to identify potential shortcomings in the process safety management (PSM) and assess the real impact of the parameter mismatches on the operational integrity (process) security. Analytics allows you to compare historical conditions, performance, models, and trends for unaccounted-for scenarios or to identify similar conditions in a project that can develop into an unsafe state if they are not manageable.

As part of the human-machine interface, contextual visual cues ("alarms") must be provided to help operators and senior managers quickly and easily understand the current situation and identify potential problems before they arise. Security managers and engineers need to know where the enterprise processes are in terms of process risk levels (i.e., if the system is functioning at a higher risk state than anticipated) and be warned in advance of possible deviating production conditions.

It is also recommended that cloud technologies be considered in some cases as a central point of data collection and as a means to reduce the cost of deploying infrastructure and security tools, applications, and data

(Haris et al., 2019; Qian, 2019). This makes it possible to provide data for analysis to experts anywhere in the world, to understand what is happening with process safety at the enterprise, and to make decision-making on security more efficient. The cloud can be used as a single place to host digital design, analysis, and security tools so that they are easily accessible to everyone, anywhere and anytime.

The use of a digital twin as a virtual model of a physical object (enterprise) serves to define, create, test, maintain, and support security systems in a virtual environment. This combination of virtual and physical worlds makes it possible to analyze data and control systems to prevent problems before they occur, prevent downtime, develop new capabilities, and even plan for the future using modeling. At the design stage, the digital twin can be used offline to run "what if" scenarios, identify potential increased levels of risk at the design stage, etc. In the online mode, the digital twin is a dynamic model of an operating asset to prevent problems before they arise, or to find the cause of the problems.

5.2.2 Aggregated model of infrastructure object

As part of the digital twin, it is necessary to build an aggregated model of the infrastructure object, which includes models of the physical and chemical processes of production, technological operations, organizational management processes, etc. At the first stage, when constructing a conceptual model M of the system S and its formalization, a model is formulated and its formal scheme is built. The main purpose of this stage is the transition from a meaningful description of the object to its mathematical model, in other words, the process of formalization (Zhijuan, Xueyan, Xianjuan, Ximing, & Bo, 2020). Computer simulation of systems is currently a universal and effective method for evaluating the characteristics of large systems. The most important and least formalized parts of the system simulation are defining the boundary between the system S and the external environment E, simplifying the description of the system, as well as building a conceptual and formal models of the system. The model must be adequate; otherwise it is impossible to obtain positive modeling results, i.e., the study of the functioning of the system on an inadequate model generally loses its meaning. By an adequate model, we mean a model that, with a certain degree of approximation at the level of understanding of the simulated system S by the developer of the model, reflects the process of its functioning in the external environment E.

Consider the transition from an informal description of an object to a block (aggregated) model. It is most rational to build a model of the system's functioning according to the block principle. In this case, three autonomous groups of units (aggregates) of such a model can be distinguished. The blocks of the first group are a simulator of the effects of the external environment *E* on the system *S*; the blocks of the second group are actually a model of the process of functioning of the studied system *S*; blocks of the third group are auxiliary and serve for the machine implementation of the blocks of the first two groups, as well as for fixing and processing the simulation results. For clarity, we introduce the idea of describing the properties of the process of functioning of the system *S*, i.e., of its conceptual model M_C, as a set of some elements conventionally depicted as a square, as shown in Fig. 5.8A.

The squares are a description of some subprocesses of functioning of the system under study *S*, the influence of the external environment *E*, etc. The

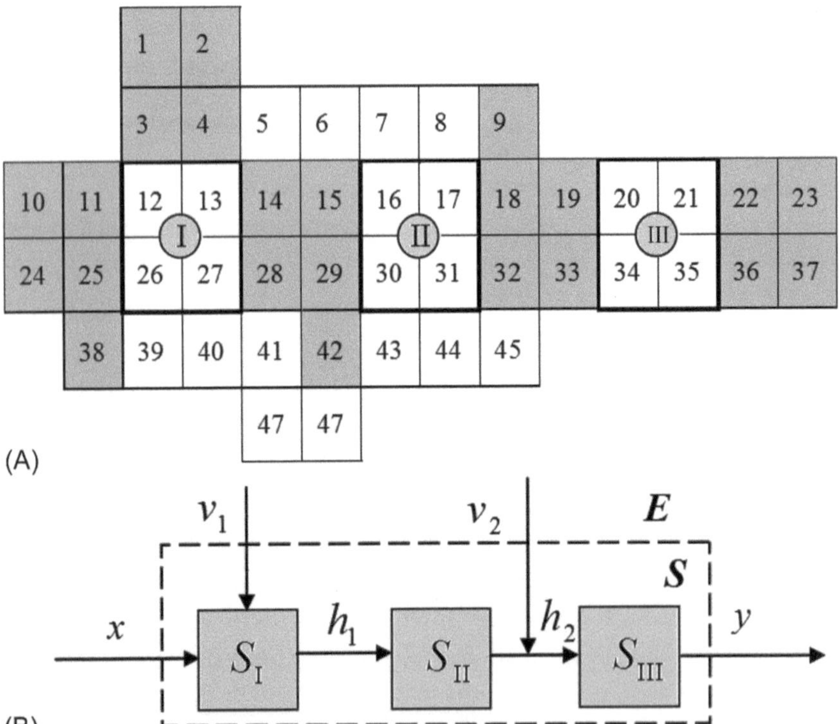

Fig. 5.8 System models: (A) conceptual; (B) aggregated.

transition from the description of the system to its formal model in this interpretation is reduced to excluding from consideration some secondary description elements (for this example, these are elements 5–8, 39–41, 43–47). It is assumed that they do not significantly affect the course of the processes studied using the model. Some of the elements (14, 15, 28, 29, 42) are replaced by passive bonds h_1, reflecting the internal properties of the system (Fig. 5.8B). Some of the elements (1–4, 10, 11, 24, 25) are replaced by input factors x and environmental influences v_1. Combined replacements are also possible: elements 9, 18, 19, 32, 33 are replaced by a passive coupling h_2 and the influence of the external environment E. Elements 22, 23, 36, 37 reflect the effect of the system on the external environment y.

The remaining elements of the system S are grouped into blocks S_{I}, S_{II}, S_{III}, reflecting the functioning of the system under study. Each of these blocks is autonomous, which is expressed in a minimum number of connections between them. The behavior of these blocks should be well studied, and for each of them a mathematical model is constructed, which in turn can also contain a number of subblocks. The constructed block (aggregated) model of the process of functioning of the studied system S is designed to analyze the characteristics of this process, which can be carried out with a computer implementation of the resulting model.

After moving from the description of the simulated system S to its model M_C, constructed according to the block principle, it is necessary to construct mathematical models of processes that occur in various blocks. The mathematical model is a set of relations (e.g., equations, logical conditions, and operators) that determine the characteristics of the process of functioning of the system S depending on the structure of the system, behavior algorithms, system parameters, environmental influences E, initial conditions, and time. The mathematical model is the result of formalizing the functioning of the system under study, i.e., constructing a formal (mathematical) description of the process with the degree of approximation to reality that is necessary in the framework of the study.

To illustrate the possibilities of formalization, we consider the functioning process of a hypothetical system S, which can be divided into m subsystems with characteristics $y_1(t)$, $y_2(t)$, …, $y_{nY}(t)$ with parameters h_1, h_2, …, h_{nH} in the presence of input x_1, x_2, …, x_{nX} and environmental influences v_1, v_2, …, v_{nV}. Then the mathematical model of the process can serve as a system of relations of the form:

$$\left.\begin{aligned} Y_1(t) &= f_1(x_1, x_2, \ldots, x_{nX}; v_1, v_2, \ldots, v_{nV}; h_1, h_2, \ldots, h_{nH}; t); \\ Y_2(t) &= f_2(x_1, x_2, \ldots, x_{nX}; v_1, v_2, \ldots, v_{nV}; h_1, h_2, \ldots, h_{nH}; t); \\ &\ldots \quad \ldots \quad \ldots \quad \ldots \\ Y_{nY}(t) &= f_m(x_1, x_2, \ldots, x_{nX}; v_1, v_2, \ldots, v_{nV}; h_1, h_2, \ldots, h_{nH}; t). \end{aligned}\right\} \tag{5.8}$$

If the functions $f_1, f_2, \ldots, f_m$ were known, then relations (5.8) would be an ideal mathematical model of the process of functioning of the system *S*. However, in practice, obtaining a model of a fairly simple form for large systems is most often impossible; therefore, usually the process of functioning of the system *S* is divided into a number of elementary subprocesses. At the same time, it is necessary to divide into subprocesses in such a way that the construction of models of individual subprocesses is elementary and does not cause difficulties in formalization. Thus, at this stage, the essence of the formalization of subprocesses will consist in the selection of typical mathematical schemes. For example, for stochastic processes, these can be schemes of probabilistic automata (*P*-schemes), queuing schemes (*Q*-schemes), etc., which quite accurately describe the main features of real phenomena that make up subprocesses from the point of view of solved applied problems (Otumba, Mwambi, & Onyango, 2012; Randhawa, Shaffer, & Tyson, 2009; Tu, Zhou, Cui, & Li, 2019).

Thus, the formalization of the functioning of any system *S* must be preceded by a study of its constituent phenomena. As a result, a meaningful description of the process appears, which is the first attempt to state clearly the patterns characteristic of the process under study, and the statement of the applied problem. A substantial description is the source material for the subsequent stages of formalization: constructing a formalized scheme of the process of functioning of the system and a mathematical model of this process. For the subsequent integration of the aggregated model into a digital twin, it is necessary to convert the mathematical model of the process into the corresponding modeling algorithm and computer program.

5.2.3 Hierarchical system of models

Digital twin technology becomes much more effective when applied to the entire fleet of production equipment—the digital fleet. The fleet combines many similar objects of the same functional purpose. The concepts of a fleet of vehicles, for example, ships, automobiles, aircraft, are widely known. You can also talk about the fleet of units for catalytic cracking of oil or the fleet of offshore oil platforms. Using the concept of a digital fleet, it is possible to conduct a comparative analysis of individual twins with each other as applied to similar equipment (units or plants) regardless of the manufacturer, owner, geographical location, or other characteristics (Monnin, Leger, & Morel, 2011). The increased amount of data increases the accuracy of the model's analytics, allowing twins to develop based on the experience of operating the entire fleet, making each individual model more accurate with more and more equipment connected to the twin. Such joint analysis within the whole fleet helps to identify common features between the equipment with the highest or lowest efficiency for potentially changing key performance indicators and opportunities to improve operation and maintenance, as well as ensuring process safety and effective risk management.

Separate plants of the fleet should have some characteristics that allow them to be grouped in accordance with a specific purpose. Considering domain-specific attributes, a systematic approach to managing data at the fleet level allows you to analyze data and information by comparing in accordance with the points of view of different heterogeneous plants (for example, compare the degree of wear of similar equipment in different places, compare different trends in system performance for similar technological processes, etc.). A system-wide approach provides a coherent structure that allows data and models to be combined to support diagnostics, forecasting, and expert knowledge through a global and structured view of the system, and increases the effectiveness of responses to emergency situations. When managing the fleet, the following problems arise:

- processing a large amount of heterogeneous monitoring data (for example, interpreted performance assessment);
- facilitating diagnostics and forecasting for a heterogeneous fleet of equipment (for example, when using different technologies and operating protocols at different enterprises); and
- providing management personnel (engineers and safety managers) with effective means of supporting decision-making throughout the entire equipment life cycle.

To solve the indicated problems, a complete instructive guide (model, method, and tool) is needed to support such processes as monitoring, diagnostics, forecasting, and decision support on a fleet scale.

(Armendia, Alzaga, & Euhus, 2019; Lee, Bagheri, & Kao, 2015).

In addition to individual modeling and monitoring of the safety of individual pieces of equipment, it is necessary to provide facilities for the entire fleet. To this end, a platform based on a hierarchical system of models can integrate semantic modeling, which allows you to collect and share data, models, and experience of various systems and equipment throughout the fleet. The fleet-based data management approach depicted in Fig. 5.9 is built on the basis of a systematic approach that makes it possible to develop and

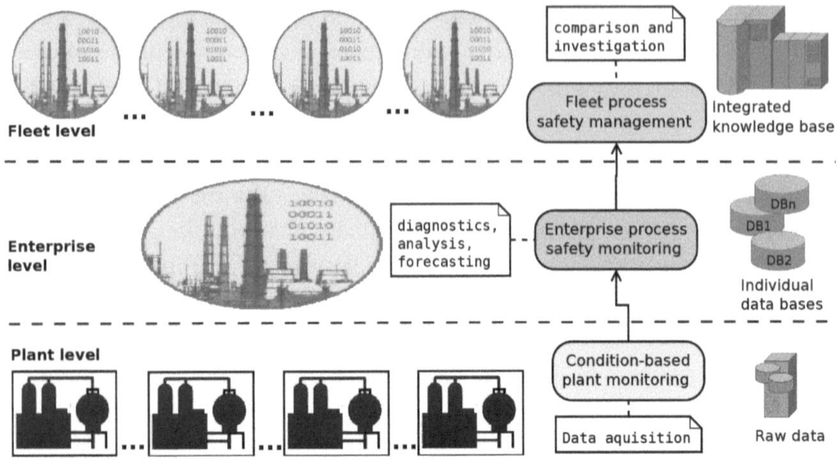

Fig. 5.9 Hierarchical system of models for digital twin fleet management.

adjust on-line strategy for managing process risks throughout the entire equipment life cycle.

This approach provides a consistent structure that facilitates combining data and models to support diagnostics, forecasting, and expert knowledge through a global and structured representation of the system with the aim of predicting a possible critical situation in advance at the level of monitoring the state of the system (fleet) as a whole.

According to this approach, the goals and key indicators of the functioning of the system, as well as its relationship with the environment, are first determined. Then the formalization of the analysis of the process and operational risks of the system is considered. Relevant knowledge, for example, models of functioning and malfunctioning, is gradually built in a knowledge-based system that supports a structured and hierarchical description of the fleet throughout its life cycle. This formalization reduces the efforts to consolidate data in the process safety management system, as well as to aggregate knowledge in fleet safety management. This is made possible by the exchange of data and knowledge between and within levels.

5.2.4 Problem of digital twin accuracy

Inaccuracy of digital twins can be a problem. The question arises as to what is the possible complete degree of accuracy of digital twins. The main and most obvious advantage of digital twin technology is its exceptional ability to create models with which you can easily simulate various technological processes, operations, and organizational control actions. This concept is beginning to become more and more common in the manufacturing sector, including in the process industry, but there are a number of factors that limit the potential of digital counterparts (Harper, Ganz, & Malakuti, 2019; Jones, Snider, Nassehi, Yon, & Hicks, 2020; Monsone & Jósvai, 2019).

The greatest concern for most owners of enterprises interested in this technology is the risk of incorrect display in the model of the object or the system that they want to reproduce with the support of this technology. The situation is complicated by the fact that the amount of information and methods of comparison—how accurate the twin is compared with its physical counterpart—is very limited. If we talk about the process of creating virtual twin objects, it seems at first glance quite simple, accurate, and understandable. The specific sensors that are used and connected in order for the interface to scan the object successfully create a very detailed picture of the physical object by its parameters. Moreover, with proper use of digital twin equipment, a model using sensors can also generate an internal latent picture of the functioning of a physical object. For example, by measuring the temperature, pressure, and flow density in the catalyst pipe of a cracking unit, one can indirectly judge the normal course of the process and predict the possible emergence of a critical situation (for example, the danger of explosion or fatigue failure of the structure).

Depending on the complexity of the physical element that needs to be reproduced, you may encounter several difficulties when it comes to creating a digital replica of the internal (hidden, poorly observable, or poorly formalized) part of the simulated object. Therefore, in order to solve this problem, the physical object may need to be open, observable, or its internal structure should be presented manually using animation and code. For example, when it comes to representing organizational subsystems and processes in a model, such as process safety management, stakeholder relationships, or staff briefing, the process may seem more complicated at first glance. In fact, it is necessary to create a clear, transparent, and accurate picture of the organizational processes that you want to digitize. In most cases, this includes working with digital twinning software, visualization in the form of flow diagrams or action diagrams, and manual verification to ensure the necessary modeling accuracy (Evangeline & Anandhakumar, 2020; Tao, Zhang, & Nee, 2019a).

Thus, a lot depends on the quality of the information received, that is, on the characteristics of the sensors, which are often combined into an industrial

internet of things (then we can talk about IIoT Digital Twin). In this case, digital models of sensors should have the following properties:
- visibility and accurate generation of all expected sensor data and system data, sizes, and speeds that will be received by the end device;
- visibility and accurate generation of all levels of protocol data, frame sizes, and frame types that will be used to send data from the network of end devices to the server;
- visibility of all data link networks, from the end device to the server;
- visibility and accurate generation of all error information generated by the system, from the end device to the server; and
- visibility and accurate generation of all security protocols.

A digital twin's accuracy increases over time as more data is refined by the model and as similar resources are deployed with their own digital twins. Data is constantly collected to maintain a modern model. The digital twin provides detailed asset knowledge, what-if forecasts, and a real-life guide from which you can create applications to optimize services, productivity, efficiency, service, supply chain, business operations, and much more.

Another challenge is the accuracy of a future digital twin simulation scenario. To explain how accurate the scenario simulations will be, we must first understand that the accuracy and degree of realism of the simulation will strictly depend on how well the work was done when creating the original digital twin model. With this in mind, the rest of the simulation will be based on the rules that were specified as part of the digital platform tools, and on the amount of data that the digital twin has accumulated from the physical object.

Thus, it must be borne in mind that all models have their limitations. Digital twins are developed using sounding, data, and modeling strategies with capabilities, system coverage, and accuracy that are achievable and required to solve specific problems. Digital twin models include all the necessary aspects of a physical object or a larger system, for example, a fleet of objects, including thermal, mechanical, electrical, chemical, hydrodynamic, material, economic, and statistical.

5.3 Simulation in real time mode

Computer modeling is a useful tool for both the design of processes and for the design of control systems in chemical production. Computer modeling, as a rule, involves the use of computers to perform various computing processes in stationary (or equilibrium) conditions. Computer models can include material, energy, and impulse balances, as well as phase equilibrium relationships, chemical reaction rate expressions, physical property correlations, equation of state, and equipment correlations (e.g., head/flow curves for compressors and pumps). Modeling is used in the process of designing a process and control equipment for stationary conditions (for example, line and valve sizes necessary for design operating speeds).

Real-time simulation is a computer model of a physical system that can run at the same speed as a simulation object in the real world. In other words, a computer model that works at the same speed as a real physical system. For example, if the actual process of distillation in the laboratory takes 35 min, then the simulation will also take 35 min.

Real-time simulation plays an important role in the process industry for training operators, setting up autonomous controllers, and managing process safety in digital twins (Chen & Büskens, 2017; Dghais, Alam, & Chen, 2018; Martínez, Cuevas, & Drake, 2012). Computer systems such as Simulink, LabVIEW, and VisSim allow you to create quickly such simulations in real time and have the ability to connect to industrial interfaces and programmable logic controllers via OLE (object linking and embedding)—technology for linking and embedding objects in other documents and objects for controlling the process or digital and analog input/output (I/O) cards.

In real-time simulation, simulation is performed at a discrete time with a constant step, also known as fixed-step simulation, when time moves forward at regular intervals. In this case, the model time required to solve the equations of the internal state and the functions representing the system should be less than a fixed step. If the calculation time exceeds the time of a fixed step, it is considered that an excess has occurred. Simply put, real-time simulation should create internal variables and output them over the same period of time as their physical counterpart. Setting up models for real-time operation allows the use of hardware modeling in a loop for testing controllers. You can make design changes earlier in the development process, reducing costs and shortening the design cycle.

Real-time simulators are widely used in many areas of technology. As a result, the inclusion of modeling applications in curricula can provide greater efficiency in the training and retraining of chemical plant operators, engineers, and safety managers. Examples of real-time simulator technology applications include statistical testing of a pipeline safety valve system, design and simulation of a catalytic cracking unit, and design methods for gas turbine plants of gas pumping stations.

Real-time simulation involves modeling the time-varying behavior of a process by solving dynamic forms of balance equations (i.e., ordinary and partial differential equations with time as an independent variable). Thus, modeling finds direct application in dynamics and process control. In particular, real-time simulation involves simulating a model on a computer that is connected to external equipment (and sometimes called a simulation in the hardware circuit). For example, in the case where a computer simulates a distillation column, the external equipment may be a basic process control system specifically configured to control the distillation column.

A basic process control system is a system that responds to inputs from a process and related equipment, other programmable systems, and/or from an operator, and generates output signals, causing the process and related equipment to work in the desired manner and within normal production constraints. Typically, a basic process control system performs the following functions:

- control of the process in predetermined working conditions, optimization of the installation for the production of a quality product, and maintenance of all technological parameters within the safe functioning of the equipment;
- providing signaling/recording events and trends;
- providing an operator interface for monitoring and control through the operator console (human machine interface); and
- reporting on production data.

Real-time computer simulations can also be used to improve the safety and efficiency of process control by providing the means to test control equipment thoroughly before actually using it in a factory. Factory acceptance tests should be carried out with realistic simulations to confirm that the equipment meets the specifications of the project and that the equipment works adequately under load conditions. Such computer simulation should be verified with mathematical or conceptual models and, if possible, again confirmed by the actual plant data.

The simulation volume can be an entire processing unit (the so-called process flow simulation) or one specific plant component. Simple simulation can be performed on most BPCS and even on single-loop controllers operating in simulation mode. However, accurate simulation of real-time control usually requires specialized high-speed computers with real-time I/O, as well as carefully selected methods of numerical integration to ensure the following conditions:

- The computer on which the simulation is performed supports synchronization with external equipment (that is, the time in the computer is not accelerating or slowing down) (Ashtari, Nasser, Wolfgang, & Michael, 2018).

- External computer input signals are not required for numerical integration calculations until they are actually read from external equipment (for example, conversion and updates in the case of analog signals).
- The sizes of the steps of numerical integration are small enough for the accuracy and stability of numbers, and also small in comparison with the control interval (or sampling time) of external (digital) control equipment.

One existing example of the application of real-time models is the model of predictive control and dynamic optimization in real time of steam cracking plants (Rossi, Rovaglio, & Manenti, 2019). This model uses two advanced model-based optimization and control strategies, called predictive model control and real-time dynamic optimization, and demonstrates how they can improve the efficiency and safety of chemical processes. As a control object, a steam cracking furnace is used.

Another example is the determination in real time of the optimal switching time for a H_2 production process with CO_2 capture using Gaussian regression models of the process (Zanella et al., 2019). This paper presents a systematic methodology for determining in real time the optimal duration of the three stages of a new Ca-Cu loop formation process for H_2 production with integrated CO_2 capture. Economic and qualitative criteria are proposed to determine the appropriate time for switching between stages. These criteria are based on time profiles of some key variables, such as product concentration. Considering the delayed nature of measurements of hardware sensors, the determination of such variables in real time is based on soft sensors. Gauss regression models are used for this.

In addition, in real time, on the basis of the model, corrosion was predicted in heat exchangers (Holzer & Wallek, 2018). This study proposes a process forecasting method for real-time evaluation of the minimum possible surface temperature of a heat exchanger, taking into account the full use of the process optimization potential. Key features of this approach include the use of strict thermodynamics, taking into account all the relevant components of the installation, which are necessary for a complete balance of mass and energy, and a strict calculation of the heat exchanger, ensuring the surface temperature and the temperature of the dead zone for the current thermodynamic state. Individual computational fluid dynamics (CFD) modeling, which is integrated into the simulation model, provides a complete spatial distribution of heat exchanger surface temperatures and dead zone temperatures. Both components are combined into a process prediction model, where the interface between the process prediction model and the process control system (PCS) is used to transmit in real time the current process parameters to the model and the recommended process parameters generated by the model back. These recommendations can either be used as operator guidelines or directly implemented as PCS command variables, taking into account the automatic optimal operating mode.

5.3.1 Edge computing

As described above, the internet of things is a network or interconnection of devices, sensors, or actuators that exchange information through a single protocol. These devices collect data on industrial equipment everywhere, analyze the data received, and provide information.

Until recently, cloud computing was considered a traditional technology that complemented the internet of things (Shi, Cao, Zhang, Li, & Xu, 2016; Tao, Zhang, & Nee, 2019b; Um, Gezer, Wagner, & Ruskowski, 2020). Cloud computing is defined as a model that provides ubiquitous and convenient access on demand to a common set of computing resources, such as storage, servers, networks, applications, and services. These resources can be quickly prepared and provided for users with minimal interaction between the control center and the service provider. Cloud computing often uses a centralized server as its primary computing resource, which is typically geographically remote from end users. This increases the frequency of data exchange between a huge number of peripheral devices used by users in enterprises (sensors, actuators), which becomes a limitation for the correct operation of applications that require real-time response.

This difficulty led to the emergence of edge computing technology, which makes it possible to process data directly on local network nodes near data sources. In edge computing, the end device not only consumes data, but also creates data (Shi & Dustdar, 2016; Shi, Pallis, & Xu, 2019; Zoualfaghari, Beddus, & Taherizadeh, 2020). A digital twin can also expand its capabilities by introducing edge or fog computing, which can be a solution to reduce connection problems and network delays (Pushpa & Kalyani, 2020; Qi, Zhao, Liao, & Tao, 2018).

In edge computing, nodes can be centralized, distributed (basic), or located at the ends of the network, in which case they are called "edges." This provides the most distributed processing of all information generated by peripheral devices. Edge computing provides data processing geographically close to end devices, such as sensors, actuators, and IoT objects.

Peripheral computing devices can be located directly at the enterprise, in sufficient proximity to industrial facilities, or at the border of the network of a communication service provider adjacent to the access network.

The concept of the cyber-physical production system and the connection of production modules to the cloud services of artificial intelligence (AI) contributes to the widespread use of edge computing in the process industry (Qi et al., 2018; Shuiguang et al., 2020).

The technology of sharing digital twins and cyber-physical systems consists in obtaining data from physical objects and sensors for measuring object and environment parameters (for example, pressure and temperature in a pipeline, flow rate, etc.), as well as in calculating and analyzing data in cyber models to control real physical objects. As a result, a closed data loop is created that forms the interaction and fusion of the physical world and cyber models at various levels of the hierarchical system of fleet management models based on digital twins (see Section 5.2.3). Boundary computing, fog computing, and cloud computing are the implementation of cyber-physical systems within the framework of digital twins at the level of production facilities, enterprise level, and digital fleet level (Fig. 5.10).

Production resources (for example, a distillation column or a catalytic cracking unit) together with a computer part (for example, an embedded system) form cyber-physical systems and a digital twin of the level of the production unit (plant). Using sensors, the cyber-model of the installation can control and perceive information from physical devices and control physical devices through an actuator (for example, a valve) capable of receiving control instructions. Due to the capabilities of perception, data analysis, and control, peripheral computing can be deployed on cyber-physical systems of the production plant level, which can be considered as border nodes. Because data is circulated directly to the device located on the production plant itself, peripheral computing can implement simpler and more compact software applications that can help provide more real-time responses.

For example, on a reduced catalyst pipe, which is part of a catalytic cracking unit, sensors are used to determine if certain deviations of the parameters pose a threat to process safety or not. In this case, any delay can have fatal consequences. If the data is sent directly to the cloud, the response time may be too long. However, in peripheral computing, the delay is significantly reduced, which allows you to make decisions in real time, since the data is not far from the data source.

Several cyber-physical systems and digital twins at the plant level are connected to the network through a network interface and information

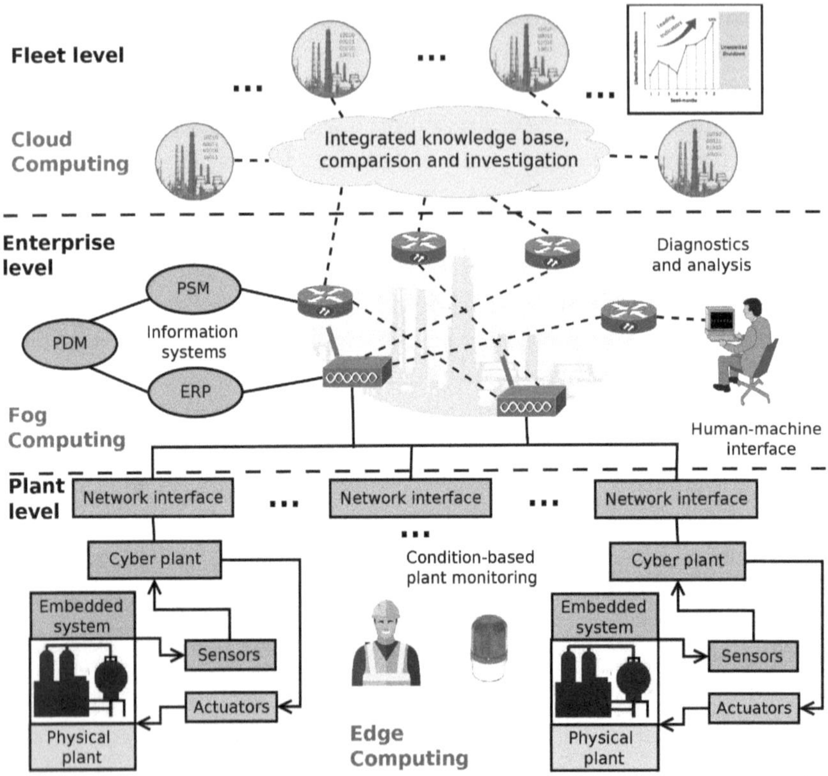

Fig. 5.10 Edge computing for digital twin fleet management.

management systems [for example, process safety management (PSM), product data management (PDM), or enterprise resource planning (ERP)]. System level cyber-physical systems combine various heterogeneous cyber-physical systems. In addition, through the human-computer interface, each digital twin of the level of the production facility can be accessed and monitored to monitor and diagnose the state and control safety of the corresponding system. As a rule, enterprise-level cyber-physical systems are geographically concentrated within a single production complex, which is well suited for the fog computing model (Hungud & Arunachalam, 2020; Tao et al., 2019b). Internal data of digital twins at the manufacturing enterprise level can be directly processed using fog computing at the nodes of the enterprise network (such as base stations, proxies, routers, and others) using intelligent data analysis applications, rather than being transmitted to the cloud and back, which is especially important in view of possible delays, network traffic, high costs, and so on.

Further, several enterprise-level cyber-physical systems can be combined into a fleet-level cyber-physical system (Fig. 5.10). For example, several geographically distributed refineries in the process industry collaborate with each other through an intelligent cloud services platform, integrating the entire system to ensure process safety throughout the entire product life cycle. The integrated data of digital twins of the fleet level become larger in volume and much more diverse, thus turning into big data. At the same time, at the fleet level, distributed storage and processing of data and knowledge, as well as the provision of data and intelligent services for a joint investigation, are required. In the cloud computing architecture, many different types of storage devices can work together with application software that makes them compatible. In addition, big data analysis must be supported by distributed processing and virtualization technologies.

Thus, cloud computing is an ideal technology for long-term and massive data storage and analysis of cyber-physical systems at the level of a fleet of digital twins. By virtue of cloud architecture, various cyber-physical systems can be encapsulated in services to become plug-and-play components and shared with other users.

At the same time, the application of edge computing technology at the lower level (industrial plant level) "unloads" the upper levels to perform complex resource-intensive "slow" tasks, such as strategy development, comparison and investigation.

5.3.2 Extreme learning machines

Recently, in the framework of digital twins, not only have models based on the mathematical representation of physical and chemical laws been actively used, but also big data technologies, such as data analysis and machine learning, together with the internet of things.

Extreme learning machines are feed forward connected neural networks. They are used for classifying, regressing, clustering, sparse approximation, and compression. These neural networks have one or several layers of hidden nodes. All the parameters of hidden nodes require tuning. These hidden nodes can be randomly assigned and never updated (that is, they are a random projection, but with nonlinear transformations) or they can be inherited from their ancestors without change. In most cases, the output weights of hidden nodes are usually learned in one step, which is actually equivalent to learning a linear model. These models are able to provide good generalization performance and learn thousands of times faster than networks trained using back propagation (Chen, Li, Duan, & Li, 2017; Huang, 2015; Huang, Zhou, Ding, & Zhang, 2012).

The presentation of neural networks in the form of a "black box" in general, and extreme learning machines in particular, is one of the main problems that limits their use in safety-critical automation tasks. One approach to solving this problem is to reduce the dependence on random input. Another approach focuses on the inclusion of permanent constraints in the learning process of extreme learning machines, which are based on prior knowledge of a particular task. Studies have shown that a special form of extreme learning machine with functional separation and linear read-out weights is especially suitable for the effective inclusion of continuous constraints in predefined areas of the input space (Huang, 2014; Neumann, 2014; Neumann, Rolf, & Steil, 2013).

The possible application of machine learning models, particularly extreme learning machines, as the key part for digital twin technologies would be based on quite different principles than the physical model-based solution.

Physics- and chemistry-based models have the following features. The models are based on the mathematical equations of the main physical and chemical laws. Sensors are used to measure physical process parameters. Decision support is based on aggregated process performance analysis. In contrast to this, machine learning models are based on statistics and data science technologies. The model is built by learning from data stream monitoring. The model is used to predict output decisions on the basis of input sensor patterns. And, finally, decision support is provided by monitoring, diagnostics and prediction (Fig. 5.11).

Consider the main advantages and disadvantages of these two approaches in the framework of digital twins. Models based on chemistry and physics make it possible to build functional cause-effect relationships in the studied object. At the same time, the struggle with the uncertainty of the processes is carried out with the help of precise input parameters and the accuracy of the model itself. However, the construction of such models requires a large amount of a priori knowledge about the object under study and overcoming computational complexity.

For models built on machine learning, a large amount of basic knowledge is not required, since they are built solely on the basis of big data. Such models are flexible and universal; they support heterogeneous data streams. Machine learning models improve over time as data accumulates, helping to identify hidden relationships between parameters and processes.

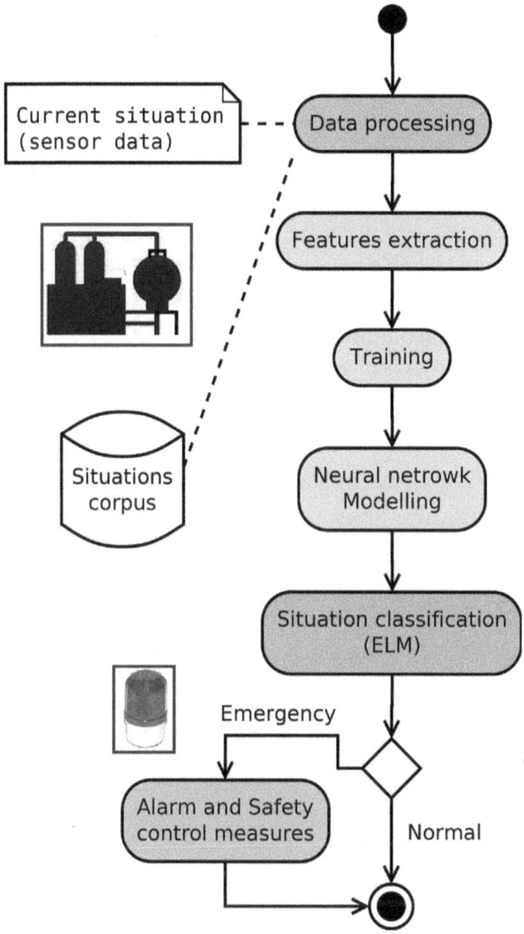

Fig. 5.11 Situations classification using extreme learning machines (ELM).

However, building a machine learning model requires a specific set of training data. The disadvantages include the fact that such models make it possible to establish the relationship, but not cause-effect relationships. In addition, approximation methods are used in the models, and not mathematics in an explicit form. Sometimes it is difficult to predict extreme conditions in a system with a small number of observations.

Thus, we can conclude that models based on physics and chemistry and models based on machine learning can complement each other within the framework of digital twins, combining the strengths of each of these technologies. Based on providing a close-to-real-time digital representation of the state of the production facility in real time, digital twin offers an overview of the functioning history of the investigated object, allowing you to both aggregate behavior and performance over time, and to conduct a retrospective analysis to understand cause-effect relationships and behavior patterns based on big data. Along with gaining an understanding of the relationship between operating conditions and production chemical plant behavior, we can also predict the plant's response to expected future scenarios. By combining extreme learning machine-based modeling with physical and chemical models of the production plant processes, we can predict in advance the occurrence of a possible critical situation that poses a threat to process safety, and also adjust the control system parameters to optimize the production plant operational characteristics.

For example, one of the provided studies is dedicated to modeling post-combustion CO_2 capture process using a bootstrap aggregated extreme learning machine. The dynamic learning machine models predict CO_2 capture rate and CO_2 capture level using the following model inputs: inlet flue gas flow rate, CO_2 concentration in inlet flue gas, pressure of flue gas, temperature of flue gas, lean solvent flow rate, monoethanolamine concentration, and temperature of lean solvent. For enhancing model reliability and accuracy, multiple extreme learning machine models were developed from bootstrap resampling replications of the original training data and combined. The developed models can be used in the optimization of CO_2 capture processes in the framework of chemical and refinery processes monitoring (Bai et al., 2016).

5.4 Big data technologies for simulation and data acquisition

As mentioned above, modeling is an effective and widespread tool for analysis and forecasting of complex systems, processes, and phenomena. At the same time, a compromise must constantly be found between the accuracy and complexity of the model. For example, Monte Carlo simulations (Section 5.1.2) should be provided with input data, store intermediate results, and filter and combine output data. Thus, the volume of the data flow has a great influence on the simulation technology, since it is necessary to use large data sets to increase model accuracy and scalability, as well as to distribute and analyze data at different stages of modeling. Thus, the combined use of simulation and big data technology can significantly increase the efficiency of modeling. Big data technologies and analytics, on the other hand, also use simulation.

5.4.1 Sharing simulation and big data technology

Some simulation methods are based on discrete-event environments due to their efficiency and scalability. Using simulation, you can solve planning problems in distributed heterogeneous computing environments, access to big data in distributed environments, and a more general parallel, distributed, and cloud software and hardware architecture. The management of virtual machines and data transfer, as well as the distribution of tasks in data centers, is associated with serious problems that require widespread use of modeling and simulation. Recently, modeling and big data technologies have been used together to build new techniques for extracting knowledge and managing large-scale and heterogeneous data collections (González-Vélez, 2019; Kołodziej, Gonzalez-Velez, & Karatza, 2017; Pop, Iacono, Gribaudo, & Kołodziej, 2016).

Big data is a new technology, the size and capabilities of which go beyond the capabilities of modern simulation tools. Datasets are heterogeneous, that is, they are produced from different sources and are large in size with high I/O speeds. The availability of big data and the ability to combine data efficiently will be extremely useful in terms of simulation. Examples of their joint use can be performance evaluation and management of cloud computing structures, the study of process safety problems, and the identification of key parameters that affect security and the prediction of critical production situations.

Data analytics is used as an important component of modeling to analyze input and output parameters, which usually require analysis of large volumes of data. Analytical applications are also used to calibrate data and evaluate unknown input simulation parameters and verify simulation results (Ayyalasomayajula, Gabriel, Lindner, & Price, 2016; Li et al., 2015). On the other hand, the role of modeling in data analytics in production can be divided into two main categories: the direct use of modeling as an application for data analysis and the use of modeling to support other data analytics applications.

Descriptive analytics determines what happened in the past or is happening now. It includes the presentation of production data in a generalized form or in the form of a request to provide meaningful information. Such an analysis basically provides various representations of the collected data, such as monitoring data from device sensors and databases, and finds patterns and trends in such data. Descriptive analytics can result in visualization of production and operating data in the form of tables, charts, and diagrams to summarize and report on trends. In this case, descriptive and visual models are used.

Diagnostic analytics identifies the causes of what happened or is happening. For example, this may include an analysis of the influence of input factors and operational strategies on the performance indicators of a production installation. Diagnostic analytics may include sensitivity analysis using a simulation model of a production system that simulates a process.

Predictive analytics determines what can happen. The focus is on evaluating performance based on planned inputs. Predictive analytics uses simulation models to evaluate performance in future periods, taking into account the current set of applicable operating strategies and sets of input data, as well as what-if scenarios.

Prescriptive analytics determines how something can be done and what the consequences of these actions will be. This type of analytics is focused on defining operational strategies and input data that will lead to the desired efficiency of the production process (for example, ensuring process safety). Prescriptive analytics using simulation models can evaluate future performance by simulating operations in accordance with potential alternative plans. These plans can be improved using optimization models. All kinds of data analytics and the main analytical algorithms as well as particularities of various models for analytics will be considered in detail in the next chapter.

In addition to analytical applications, modeling can also be used as a data generator. Testing a data analytics application requires large sets of source data. It is usually difficult to find process industry enterprises that are willing to provide access to their manufacturing facilities to collect large sets of real data. Proven simulation models of real production systems can be considered as virtual enterprises that can be used to generate data in formats identical to real sensors on a real production installation. The virtual factory must be configured to provide flexibility in modeling a wide range of enterprise configurations and the required level of detail. For example, for analytics at the plant level, it should be able to simulate the physics and chemistry of the process in order to simulate data streams on temperature, pressure, forces, vibrations, flow rates, and the influence of these parameters on product quality and process safety. At a higher level of abstraction, the virtual factory should be able to generate data flows about the flow of materials, the use of resources, and shipments of products to feed data into analytical applications at the enterprise level.

The virtual factory offers ready-made basic parameter distributions. The virtual factory output streams are based on input distributions for a number of factors. Data analytics applications can then be used to collect data flows and identify changes in the main factors affecting the performance of an enterprise. The distributions established by the data analytics applications based on the output streams can be compared with the known distribution of real input data in order to evaluate the quality of the analytics.

5.4.2 Examples of simulation implementation

Let us consider few examples of implementation of simulation for data analytics in various fields, mostly in process industry, ecology, and information technologies.

Example 5.4.2.1 (Xiaowei & Jianxi, 2019)

The research considers the task of studying the exact spatial distribution of carbon emissions in order to develop carbon reduction strategies. Previous studies had a fairly low spatial resolution of the models used and, as a consequence, low accuracy. In contrast, the approach under consideration allowed the authors to combine different sources of big data and develop a new methodology for studying carbon emissions in the city of Nanjing with a high resolution of 300m. In addition, regional differences and factors of influence were compared. This study showed that in the central part of the city there was an obvious variation in intensity, but the emission intensity was much lower than in the peripheral area, where industrial zones were mainly distributed. At the regional level, districts in the central part of the city have always had a high emission rate. The characteristics of land use and socioeconomic development were key factors in determining carbon emissions. Increasing the number of green zones and reducing the number of developed zones will greatly help carbon reduction strategies. Thus, as part of the study, it was shown that the socioeconomic development of the region and the adjustment of the structure of industrial zones, as well as improving energy efficiency, will play a key role in reducing carbon emissions.

Example 5.4.2.2 (Küçükkeçeci & Yazıcı, 2018)

This study addresses the problem of analyzing large multimedia data based on high-level modeling. The data sources are sensors that are present in various forms around the world, such as mobile phones, surveillance cameras, smart TVs, smart refrigerators, etc. Typically, most sensors are part of some other system with similar sensors that make up the network. One of these networks consists of millions of sensors connected to the internet (IoT). Many studies have already been conducted in wireless multimedia sensor networks in various fields, such as fire detection, city surveillance, early warning systems, etc. All these applications position sensor nodes and collect their data for a long period of time with data flow in real time, which is considered big data. Big data can be structured or unstructured and must be stored for further processing and analysis. The analysis of large multimedia data is a complex task requiring high-level modeling to extract efficiently valuable information/knowledge from data. This study proposes a large database model based on a graph database model for processing data generated by wireless multimedia sensor networks. A simulator for generating synthetic data and storing and querying big data using a graph model as a large database is presented. For this purpose, the well-known graph-based NoSQL, Neo4j, and OrientDB databases, and the relational MySQL database were considered. A series of experiments were conducted with the queries developed by the simulator to show which database system for monitoring in wireless multimedia sensor networks is the most efficient and scalable.

Example 5.4.2.3 (Jin, Hyeon-Ju, & Kim, 2015)

The research proposes big data analysis of hollow fiber direct contact membrane distillation (HFDCMD) for simulation-based empirical analysis. Often, HFDCMD is modeled using the previously developed corresponding module of the environmental software package (EnPhySoft). Of the 11,059,200 cases, 7,453,717 cases were physically significant for practical use. Self-organizing map (SOM) and multiple linear regression (MLR) methods were used for statistical analysis of big data. Using raw data, physical and dimensionless data sets were prepared in specific formats: the first determines the most significant parameters, and the second compares the relative importance of the input parameters. SOM analysis did not provide transparent dependencies between the inputs and/or outputs of the HFDCMD; instead, it helped to distribute the parameters into groups of similar characteristics. Using MLR, we found that macroscopic values, such as temperature and radii of the lumen and sides of the shell, had a greater effect on the characteristics of MD than microscopic values, such as pore size and membrane length. Thus, it was found that a rough forecast for heat and mass fluxes requires only four main input parameters.

Example 5.4.2.4 (Jean-Pierre, Stéphane, Fabrice, David, & Benoît, 2014)

This example considers the collaborative simulation and scientific big data analysis implementation for sustainability in natural hazards management and chemical process engineering. The classic approaches to remote visualization and collaboration used in CAD/CAE applications are no longer suitable due to the growing volume of generated data, especially using standard networks. As part of the study, an easy computing platform for scientific modeling, collaboration in the field of design, 3D-visualization, and big data management is proposed. This ICT-based platform provides scientists with an "easy-to-integrate" universal tool that enables collaboration around the world and remote processing of any data format. Service-oriented architecture is based on the cloud computing paradigm and is based on standard internet technologies that are effective for a large group of networks and customers. This platform is available as an open-source project with all the main components licensed under LGPL V2.1. As an example of the use of the platform, chemical technological design is considered, showing the development of services specific to a particular area. With the emergence of global warming problems and the growing importance of sustainable development, chemical technology requires an increasing attention to ecology. Indeed, the chemical engineer must now take into account not only the engineering and economic criteria of the process, but also its environmental and social characteristics. Secondly, the disaster management example illustrates the effectiveness of the proposed approach to remote collaboration, which includes the exchange of big data and analysis between remote locations.

5.5 Summary

Modeling is one of the main methods for studying complex organizational and technical systems in various fields of human activity, including in the rapidly developing information field and process industry and process safety. IEC 31010:2019 recommends using a modeling including software models to analyze collected information for the process risk assessment. It can be used to understand the meaning of data and to simulate what can happen in reality under various conditions. Due to the great complexity of objects in process industries, models can also have a complex hierarchical structure, often built into the framework of digital twins. When implementing such complex models, taking into account uncertainties, the accuracy of the results is of great importance. The following groups of errors can be set: modeling errors arising due to inaccurate input data; modeling errors resulting from the simplification of the original simulation model; errors in the calculation of state variables and model output parameters due to discrete implementation of the simulation model; and modeling errors due to the limited amount of statistical data or a limited number of random model tests on a computer. Various algorithms based on probability theory and mathematical statistics can be used to implement the developed models (computer simulation) of the process industry as part of digital twins using big data. Monte Carlo analysis is widely used to model input random influences (source data), and Markov models are used to study the processes of the functioning of the system. These two methods were already mentioned in the second chapter when considering the IEC 31010:2109 standard as recommended techniques for use in risk analysis, including process safety risk.

The use of digital twins is a modern and promising area of sharing modeling and big data technologies within the framework of the Industry 4.0 concept. For the process industry and process safety, the aim of the digital twin in this area is the early detection of deviations of production process parameters from the set, prediction and prevention of critical situations, support for decision-making on process safety management (PSM), production of better products, risk reduction, and safety of people and the environment. A digital twin can be an integrated model of an already created production facility. This makes it possible to simulate the behavior of an object in real time. It is also recommended that cloud technologies be considered in some cases as a central point of data collection and as a means to reduce the cost of deploying infrastructure and security tools, applications, and data. As part of the digital twin, it is necessary to build an aggregated model of the infrastructure object, which includes models of the physical and chemical processes of production, technological operations, organizational management processes, etc.

Digital twin technology becomes much more effective when applied to the entire fleet of production equipment—the digital fleet. The fleet combines many similar objects of the same functional purpose. The increased amount of data increases the accuracy of the model's analytics, allowing twins to develop based on the experience of operating the entire fleet, making each individual model more accurate with more and more equipment connected to the twin.

One of the problems of digital twins is inaccuracy. The greatest concern for most owners of enterprises interested in this technology is the risk of incorrect display in the model of the object or the system that they want to reproduce with the support of this technology. The situation is complicated by the fact that the amount of information and methods of comparison—how accurate the twin is compared with its physical counterpart—is very limited. Depending on the complexity of the physical element that needs to be reproduced, several difficulties may be encountered when it comes to creating a digital replica of the internal (hidden, poorly observable, or poorly formalized) part of the simulated object. Therefore, in order to solve this problem, the physical object may need to be open, observable, or its internal structure should be presented manually using animation and code.

Real-time simulation plays an important role in the process industry for training operators, setting up autonomous controllers, and managing process safety in digital twins. Real-time simulators are widely used in many areas of technology. As a result, the inclusion of modeling applications in curricula can provide greater efficiency in the training and retraining of chemical plant operators, engineers, and safety managers.

Digital twin can also expand its capabilities by introducing edge or fog computing, which can be a solution to reduce connection problems and network delays. In edge computing, nodes can be centralized, distributed (basic), or located at the ends of the network, in which case they are called "edges."

Today, in the framework of digital twins, not only models based on the mathematical representation of physical and chemical laws are actively used, but also big data technologies, such as data analysis and machine learning, together with the internet of things. The possible application of machine learning models, particularly extreme learning machines as the key part of digital twin technologies, would be based on quite different principles than the physical model-based solution. Data analytics is used as an important component of modeling to analyze input and output parameters, which usually require analysis of large volumes of data. Analytical applications are also used to calibrate data, evaluate unknown input simulation parameters, and verify simulation results. On the other hand, the role of modeling in data analytics in production can be divided into two main categories: the direct use of modeling as an application for data analysis and the use of modeling to support other data analytics applications.

5.6 Definitions

Modeling is a method of cognition, in which the studied original object is replaced by another model object, and the model reflects those characteristics of the object that are significant from the point of view of a particular cognitive process.

Simulation is a step-by-step reproduction of the process of functioning of the system in time.

Accuracy is one metric for evaluating classification models. Informally, accuracy is the fraction of predictions the model gets right.

Monte Carlo methods, or Monte Carlo experiments, are a broad class of computational algorithms that rely on repeated random sampling to obtain numerical results.

A Markov model is a stochastic model used to model randomly changing systems where it is assumed that future states depend only on the current state not on the events that occurred before it (that is, it assumes the Markov property).

A cyber-physical system is an information technology system that implies the integration of computing resources into physical objects of any nature, including both biological and artificially created objects.

Digital twin is a digital (virtual) copy of a physical (real) system or process created to optimize system performance.

Digital fleet is paradigm providing a comparative analysis of individual twins with each other as applied to similar equipment (units or plants) regardless of the manufacturer, owner, geographical location, or other characteristics.

Real-time simulation is a computer model of a physical system that can run at the same speed as a simulation object in the real world.

Edge computing is a distributed computing paradigm that brings computation and data storage closer to the location where it is needed, to improve response times and save bandwidth.

Extreme learning machines are feed forward connected neural networks for classifying, regressing, clustering, sparse approximation, compression and learning functions with one layer or several layers of hidden nodes, for which parameters of hidden nodes require tuning (and not just weights connecting inputs to hidden nodes).

References

Armendia, Alzaga, Peysson, & Euhus, D. (2019). *Twin-control approach: A digital twin approach to improve machine tools lifecycle*. https://doi.org/10.1007/978-3-030-02203-7_23.

Ashtari, T. B., Nasser, J., Wolfgang, S., & Michael, W. (2018). Consistency check to synchronize the digital twin of manufacturing automation based on anchor points. *Procedia CIRP*, 159–164. https://doi.org/10.1016/j.procir.2018.03.166.

Ayyalasomayajula, H., Gabriel, E., Lindner, P., & Price, D. (2016). Air quality simulations using big data programming models. In *Proceedings—2016 IEEE 2nd international conference on big data computing service and applications, BigDataService 2016*. United States: Institute of Electrical and Electronics Engineers Inc.. https://doi.org/10.1109/BigDataService.2016.26.

Bai, Z., Li, F., Zhang, J., Oko, E., Wang, M., Xiong, Z., & Huang, D. (2016). Modelling of a post-combustion CO_2 capture process using bootstrap aggregated extreme learning machines. In *Vol. 38. Computer aided chemical engineering* (pp. 2007–2012). United Kingdom: Elsevier B.V.. https://doi.org/10.1016/B978-0-444-63428-3.50339-8.

Carpenter, C. (2020). Model error estimation improves forecasting. *Journal of Petroleum Technology*, 67–68. https://doi.org/10.2118/0420-0067-jpt.

Chen, C., Li, K., Duan, M., & Li, K. (2017). Extreme learning machine and its applications in big data processing. In *Big data analytics for sensor-network collected intelligence* (pp. 117–150). China: Elsevier Inc.. https://doi.org/10.1016/B978-0-12-809393-1.00006-4.

Chen, S., & Büskens, C. (2017). Real-time model adaptation. *PAMM*, 837–838. https://doi.org/10.1002/pamm.201710386.

Chung, K. L. (2014). *A course in probability theory*. Academic Press.

Dghais, W., Alam, M., & Chen, Y. (2018). Real time modelling and processing. In *Vol. 29. Lecture notes in networks and systems* (pp. 1–13). Tunisia: Springer. https://doi.org/10.1007/978-3-319-72215-3_1.

El Saddik, A. (2018). Digital twins: The convergence of multimedia technologies. *IEEE Multimedia, 25*(2), 87–92. https://doi.org/10.1109/MMUL.2018.023121167.

Erboz, G. (2017). How to define industry 4.0: Main pillars of industry 4.0. In *7th international conference on management*.

Evangeline, P., & Anandhakumar. (2020). Digital twin technology for "smart manufacturing". In *Vol. 117. Advances in computers* (pp. 35–49). India: Academic Press Inc.. https://doi.org/10.1016/bs.adcom.2019.10.009.

Gagniuc, P. A. (2017). *Markov chains: From theory to implementation and experimentation*. John Wiley & Sons.

Galdi, P., & Tagliaferri, R. (2019). Data mining: Accuracy and error measures for classification and prediction. In S. Ranganathan, M. Gribskov, K. Nakai, & C. Schönbach (Eds.), *Encyclopedia of bioinformatics and computational biology* (pp. 431–436). Academic Press. https://doi.org/10.1016/b978-0-12-809633-8.20474-3.

Gamerman, D., & Lopes, H. F. (2006). *Stochastic simulation for Bayesian inference*. (Undefined).

Gonzaga, C. S. B., Arinelli, L.d. O., de Medeiros, J. L., & Araújo, O.d. Q. F. (2019). Automatized Monte-Carlo analysis of offshore processing of CO_2-rich natural gas: Conventional versus supersonic separator routes. *Journal of Natural Gas Science and Engineering, 69*. https://doi.org/10.1016/j.jngse.2019.102943.

González-Vélez, J. K. (2019). *High-performance modelling and simulation for big data applications*. https://doi.org/10.1007/978-3-030-16272-6.

Gronau, N., Grum, M., & Bender, B. (2016). Determining the optimal level of autonomy in cyber-physical production systems. In *IEEE international conference on industrial informatics (INDIN)*. Germany: Institute of Electrical and Electronics Engineers Inc.. https://doi.org/10.1109/INDIN.2016.7819367.

Hajek, B. (2015). *Random processes for engineers*. Cambridge University Press.

Haris, P., Bhanu, S., Singh, K. S., Sanjay, M., Sadegh, A. M., Harshit, M., … Udit, J. (2019). Transformative effects of IoT, blockchain and artificial intelligence on cloud computing: Evolution, vision, trends and open challenges. *Internet of Things*, 100118. https://doi.org/10.1016/j.iot.2019.100118.

Harper, K. E., Ganz, C., & Malakuti, S. (2019). Digital Twin Architecture and Standards. *IIC Journal of Innovation*, (November 2019), 1–8.

Hartmann, F. (2012). *Modeling error* (pp. 241–317). Springer Science and Business Media LLC.

Hastings, W. K. (1970). Monte Carlo sampling methods using Markov chains and their applications. *Biometrika, 57*(1), 97–109. https://doi.org/10.1093/biomet/57.1.97.

Holzer, G., & Wallek, T. (2018). Model-based real-time prediction of corrosion in heat exchangers. In *Vol. 43. Computer aided chemical engineering* (pp. 1255–1256). United States: Elsevier B.V.. https://doi.org/10.1016/B978-0-444-64235-6.50218-7.

Huang, G. B. (2014). An insight into extreme learning machines: Random neurons, random features and kernels. *Cognitive Computation*, *6*(3), 376–390. https://doi.org/10.1007/s12559-014-9255-2.

Huang, G. B. (2015). What are extreme learning machines? Filling the gap between Frank Rosenblatt's dream and John von Neumann's puzzle. *Cognitive Computation*, 7(3), 263–278. https://doi.org/10.1007/s12559-015-9333-0.

Huang, G.-B., Zhou, H., Ding, X., & Zhang, R. (2012). Extreme learning machine for regression and multiclass classification. *IEEE Transactions on Systems, Man, and Cybernetics—Part B: Cybernetics*, *42*(2), 513–529. https://doi.org/10.1109/tsmcb.2011.2168604. PMID 21984515.

Hungud, V., & Arunachalam, S. K. (2020). Digital twin: Empowering edge devices to be intelligent. In *Vol. 117. Advances in computers* (pp. 107–127). India: Academic Press Inc.. https://doi.org/10.1016/bs.adcom.2019.10.005.

IEC 61165:2006 Application of Markov techniques. (2006)..

ISO/IEC Guide 98-3/Suppl 1: Uncertainty of measurement—Part 3: Guide to the expression of uncertainty in measurement (GUM 1995)—Propagation of distributions using a Monte Carlo method2008. (2008)..

Jean-Pierre, B., Stéphane, N., Fabrice, D., David, M., & Benoît, V. (2014). Collaborative simulation and scientific big data analysis: Illustration for sustainability in natural hazards management and chemical process engineering. *Computers in Industry*, 521–535. https://doi.org/10.1016/j.compind.2014.01.009.

Jin, K. S., Hyeon-Ju, K., & Kim, A. S. (2015). Big data analysis of hollow fiber direct contact membrane distillation (HFDCMD) for simulation-based empirical analysis. *Desalination*, 56–67. https://doi.org/10.1016/j.desal.2014.10.008.

Jones, D., Snider, C., Nassehi, A., Yon, J., & Hicks, B. (2020). Characterising the digital twin: A systematic literature review. *CIRP Journal of Manufacturing Science and Technology*. https://doi.org/10.1016/j.cirpj.2020.02.002.

Karlin, S., & Taylor, H. M. (1975). *A first course in stochastic processes*. Academic Press. https://doi.org/10.1016/C2009-1-28569-8.

Kołodziej, J., Gonzalez-Velez, H., & Karatza, H. (2017). High-performance modelling and simulation for big data applications. *Simulation Modelling Practice and Theory*, *76*. https://doi.org/10.1016/j.simpat.2017.04.003.

Kroese, D. P., Brereton, T., Taimre, T., & Botev, Z. I. (2014). Why the Monte Carlo method is so important today. *Wiley Interdisciplinary Reviews: Computational Statistics*, *6*(6), 386–392. https://doi.org/10.1002/wics.1314.

Küçükkeçeci, C., & Yazıcı, A. (2018). Big data model simulation on a graph database for surveillance in wireless multimedia sensor networks. *Big Data Research*, 2214-5796. *11*, 33–43. https://doi.org/10.1016/j.bdr.2017.09.003. 2018.

Lee, E. A., & Seshia, E. A. (2011). *Introduction to embedded systems—A cyber-physical systems approach*. LeeSeshia.org.

Lee, J., Bagheri, B., & Kao, H.-A. (2015). A cyber-physical systems architecture for industry 4.0-based manufacturing systems. *Manufacturing Letters*, *3*, 18–23.

Li, B., Chai, X., Li, T., Hou, B., Qin, D., Zhao, Q., … Yang, M. (2015). Research on high-performance modeling and simulation for complex systems. In *Concepts and methodologies for modeling and simulation* (pp. 45–66). Springer Science and Business Media LLC. https://doi.org/10.1007/978-3-319-15096-3_3.

Manson, S. M. (2020). Simulation modeling. In A. Kobayashi (Ed.), *International encyclopedia of human geography* (2nd ed., pp. 207–212). Elsevier. https://doi.org/10.1016/B978-0-08-102295-5.10423-8.

Martínez, P. L., Cuevas, C., & Drake, J. M. (2012). Compositional real-time models. *Journal of Systems Architecture*, *58*(6–7), 257–276. https://doi.org/10.1016/j.sysarc.2012.04.001.

Meyn, S., & Tweedie, R. L. (2009). *Markov chains and stochastic stability* (2nd ed., pp. 1–504). United States: Cambridge University Press.

Monnin, M., Leger, J.-B., & Morel, D. (2011). Proactive facility fleet/plant monitoring and management. In *Proceedings of 24th international congress on condition monitoring and diagnostics engineering management, Stavanger, Norway*.

Monsone, C. R., & Jósvai, J. (2019). Challenges of digital twin system. *Acta Technica Jaurinensis*, 252–267. https://doi.org/10.14513/actatechjaur.v12.n3.514.

Neumann, K. (2014). *Reliability* (pp. 49–74). University Library Bielefeld.

Neumann, K., Rolf, M., & Steil, J. J. (2013). Reliable integration of continuous constraints into extreme learning machines. *International Journal of Uncertainty, Fuzziness and Knowledge-Based Systems*, 35–50. https://doi.org/10.1142/s021848851340014x.

Otumba, E., Mwambi, H., & Onyango, F. (2012). An aggregated model for optimal harvesting of fish. *International Journal of Ecological Economics & Statistics (IJEES)*, *27*(4).

Pop, F., Iacono, M., Gribaudo, M., & Kołodziej, J. (2016). Advances in modelling and simulation for big-data applications (AMSBA). *Concurrency and Computation: Practice and Experience*, 291–293. https://doi.org/10.1002/cpe.3750.

Pushpa, J., & Kalyani, S. A. (2020). The fog computing/edge computing to leverage digital twin. In *Vol. 117. Advances in computers* (pp. 51–77). India: Academic Press Inc.. https://doi.org/10.1016/bs.adcom.2019.09.003.

Qi, Q., Zhao, D., Liao, T. W., & Tao, F. (2018). Modeling of cyber-physical systems and digital twin based on edge computing, fog computing and cloud computing towards smart manufacturing. In *ASME 2018 13th international manufacturing science and engineering conference, MSEC 2018*. China: American Society of Mechanical Engineers (ASME). https://doi.org/10.1115/MSEC2018-6435.

Qian, W. (2019). Automatic checks from 3D point cloud data for safety regulation compliance for scaffold work platforms. *Automation in Construction*, 38–51. https://doi.org/10.1016/j.autcon.2019.04.008.

Randhawa, R., Shaffer, C. A., & Tyson, J. J. (2009). Model aggregation: A building-block approach to creating large macromolecular regulatory networks. *Bioinformatics*, *25*(24), 3289–3295. https://doi.org/10.1093/bioinformatics/btp581.

Rossi, F., Rovaglio, M., & Manenti, F. (2019). Model predictive control and dynamic real-time optimization of steam cracking units. In *Vol. 45. Computer aided chemical engineering* (pp. 873–897). United States: Elsevier B.V.. https://doi.org/10.1016/B978-0-444-64087-1.00018-8.

Saeid, M., Poe, W. A., & Mak, J. Y. (2019). Process modeling and simulation of gas processing plants. In S. Mokhatab, W. A. Poe, & J. Y. Mak (Eds.), *Handbook of natural gas transmission and processing* (4th ed., pp. 579–614). Elsevier BV. https://doi.org/10.1016/b978-0-12-815817-3.00019-8 (chapter 19).

Salimi, F., & Salimi, F. (2018). Modeling and simulation: The essential tools to manage the complexities. In F. Salimi, & F. Salimi (Eds.), *Systems approach to managing the complexities of process industries* (pp. 279–407). Elsevier BV. https://doi.org/10.1016/b978-0-12-804213-7.00005-0 (chapter 5).

Sanfelice, R. G. (2016). Analysis and design of cyber-physical systems. A hybrid control systems approach. In D. Rawat, J. Rodrigues, & I. Stojmenovic (Eds.), *Cyber-physical systems: From theory to practice* Taylor & Francis.

Shi, W., Cao, J., Zhang, Q., Li, Y., & Xu, L. (2016). Edge computing: Vision and challenges. *IEEE Internet of Things Journal*, *3*(5), 637–646. https://doi.org/10.1109/JIOT.2016.2579198.

Shi, W., & Dustdar, S. (2016). The promise of edge computing. *Computer*, *49*, 78–81. https://doi.org/10.1109/MC.2016.145.

Shi, W., Pallis, G., & Xu, Z. (2019). Edge computing. *Proceedings of the IEEE*, *107*, 1474–1481. https://doi.org/10.1109/JPROC.2019.2928287.

Shuiguang, D., Hailiang, Z., Weijia, F., Jianwei, Y., Schahram, D., & Zomaya, A. Y. (2020). Edge intelligence: The confluence of edge computing and artificial intelligence. *IEEE Internet of Things Journal*, 1. https://doi.org/10.1109/jiot.2020.2984887.

Tao, F., Zhang, M., & Nee, A. Y. C. (2019). Digital twin-driven prognostics and health management. In *Digital twin driven smart manufacturing* (pp. 141–167). Elsevier BV. https://doi.org/10.1016/b978-0-12-817630-6.00007-2 (chapter 7).

Tao, F., Zhang, M., & Nee, A. Y. C. (2019). Digital twin and cloud, fog, edge computing. In *Digital twin driven smart manufacturing* (pp. 171–181). Elsevier BV. https://doi.org/10.1016/b978-0-12-817630-6.00008-4 (chapter 8).

Tao, F., et al. (2018). Digital twin-driven product design framework. *International Journal of Production Research*, (1), 1–19.

Tu, J., Zhou, M., Cui, H., & Li, F. (2019). An equivalent aggregated model of large-scale flexible loads for load scheduling. *IEEE Access*, 7, 143431–143444. https://doi.org/10.1109/ACCESS.2019.2944233.

Um, J., Gezer, V., Wagner, A., & Ruskowski, M. (2020). Edge computing in smart production. In *Advances in intelligent systems and computing*. Germany: Springer Verlag. https://doi.org/10.1007/978-3-030-19648-6_17.

Wang, H., Wassan, J., & Zheng, H. (2018). Measurements of accuracy in biostatistics. In *Vol. 1–3. Encyclopedia of bioinformatics and computational biology: ABC of bioinformatics* (pp. 685–690). United Kingdom: Elsevier. https://doi.org/10.1016/B978-0-12-809633-8.20355-5.

Xiaowei, C., & Jianxi, F. (2019). High resolution carbon emissions simulation and spatial heterogeneity analysis based on big data in Nanjing City, China. *Science of The Total Environment*, 828–837. https://doi.org/10.1016/j.scitotenv.2019.05.138.

Yi, C., Binil, S., Paul, C., & Yuan-Shin, L. (2017). Sensor data and information fusion to construct digital-twins virtual machine tools for cyber-physical manufacturing. *Procedia Manufacturing*, 1031–1042. https://doi.org/10.1016/j.promfg.2017.07.094.

Zanella, L., Porru, M., Bottegal, G., Gallucci, F., van Sint Annaland, M., & Özkan, L. (2019). Real-time determination of optimal switching times for a H_2 production process with CO_2 capture using Gaussian Process Regression models. In Computer aided chemical engineering (Vol. 46, pp. 1219–1224). Italy: Elsevier B.V. doi:https://doi.org/10.1016/B978-0-12-818634-3.50204-6.

Zhang, L., Zeigler, B. P., & Laili, Y. (2019). Introduction to model engineering for simulation. In *Model engineering for simulation* (pp. 1–23). https://doi.org/10.1016/B978-0-12-813543-3.00001-9.

Zhang, Z., Wu, Z., Durand, H., Albalawib, F., & Christofides, P. D. (2018). On integration of model predictive control with safety system. In M. R. Eden, M. G. Ierapetritou, & G. P. Towler (Eds.), *Vol. 44. Computer aided chemical engineering* (pp. 2011–2016). Elsevier.

Zhijuan, X., Xueyan, L., Xianjuan, C., Ximing, L., & Bo, Y. (2020). Robust stochastic block model. *Neurocomputing*, 398–412. https://doi.org/10.1016/j.neucom.2019.10.069.

Zoualfaghari, M., Beddus, S., & Taherizadeh, S. (2020). Edge computing. *Computing*. https://doi.org/10.1002/9781119545293.ch3.

CHAPTER 6

Big data analytics and process safety

6.1 Analytics basics

Data analytics is an important part of the risk management system in which data is organized, processed, and analyzed, and the extraction and preparation of analytical data are presented in the form of graphs, charts, and diagrams. This requires deep knowledge of databases such as SQL (structures query language) and programming skills such as Python/R and Hadoop/Spark. It also requires knowledge of business intelligence (BI) tools and an average level of understanding of statistics. This is mainly structured data, to which the design principles and data visualization methods are applied. The scope is limited to analytical methods, mainly using statistical tools and methods.

As already noted in Section 5.4.1, there are four main types of analytics: descriptive, diagnostic, predictive, and prescriptive. These types can be attributed to different levels of understanding, assessment, and management of the state of the control object (in our case, process safety). In addition, as the types of analytics become more complicated, there is a transition from direct information and raw data to aggregated data, and, finally, to knowledge (Chan & Tang, 2019; Gudivada, Irfan, Fathi, & Rao, 2016; Nagel & Ludwig, 2019; Samson, 2019).

At the lower level, the level of descriptive analytics (and, accordingly, the initial unstructured information), a review of the current situation takes place. All data processing is limited to generating reports (basic, ad hoc, dynamic), and the results of analytics are answers to the questions "What happened?", "How much, how often and where?", and "Where exactly did the problems arise?".

At the level of diagnostic (basic) analytics, there is a gradual transition from data preprocessing to information. It is supposed to generate reports with early warnings and basic statistical analysis. Analytics prepares answers to more complex questions such as "What actions need to be taken?" or "Why is this happening?".

Process Safety and Big Data
https://doi.org/10.1016/B978-0-12-822066-5.00001-7

At an advanced level of analytics, aggregated data and knowledge are processed and intelligence is applied. This is where predictive and prescriptive analytics takes place with the goal of forecasting, predictive modeling, and decision optimization. Analysts will attempt to get answers to the questions "What will happen if these trends continue to develop?", "What will happen next?", and "What is the best solution?".

In accordance with the hierarchical model system discussed earlier (Section 5.2.3), descriptive analytics corresponds to the level of the manufacturing enterprise in this hierarchy. The reports on the diagnostics of the current situation are formed by the personnel immediately after the inspection of production equipment, which makes it possible to promptly respond to threats to process safety. Diagnostic analytics is carried out at the enterprise level. At the same time, analytical modules can be integrated into the enterprise information system. In this case, we are talking about building fairly simple predictive models of the functioning of production equipment within a single enterprise. At the fleet level (the highest hierarchy level), predictive and prescriptive analytics are performed. It is assumed that, at this level, data and knowledge regarding the process safety of a whole fleet of process industry enterprises has been collected and appropriately aggregated and processed. An illustration of the correspondence of various types of analytics to the levels of the hierarchical system of models within the framework of constructing digital twins is presented in Fig. 6.1.

Of course, the fundamental concept of analytics is data and operations with them. In Fig. 6.2, a generalized set of basic steps of the data analytics

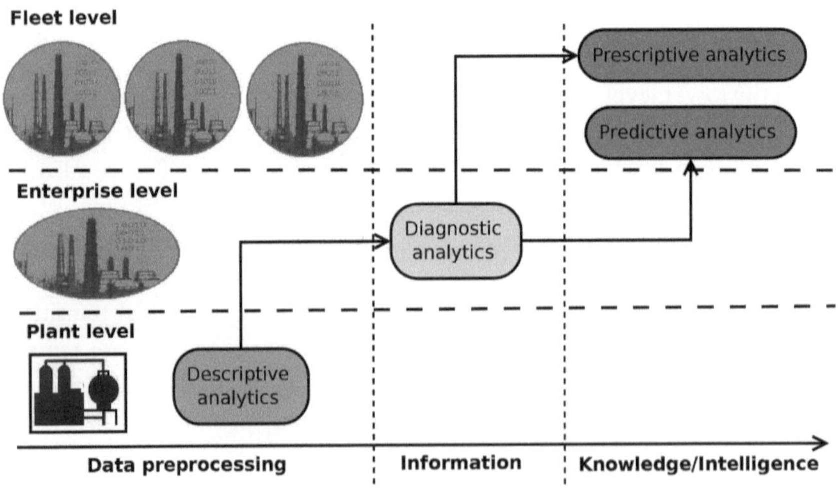

Fig. 6.1 Correspondence of various types of analytics to the levels of the hierarchical system of models in the framework of digital twins.

process are shown. The data collected are usually raw data and require pre-processing, purification, and analysis to obtain the required information for decision-making. As can be seen from the figure, the process is iterative. It is assumed that certain steps will be repeated as the data are refined and analytical algorithms become more complex.

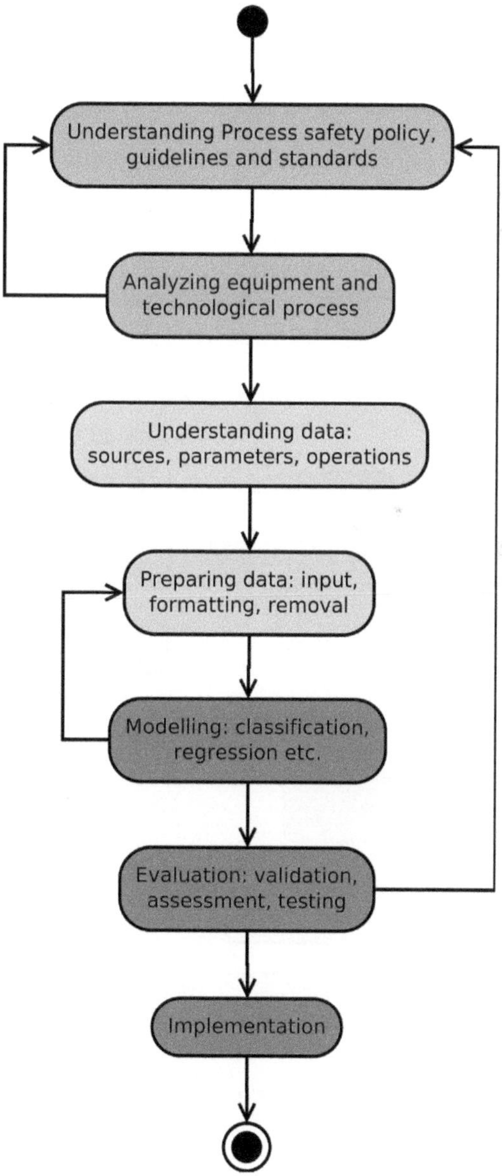

Fig. 6.2 Key steps in the data analytics process.

At the initial stage, there is an understanding of the process safety policy, guidelines, and standards in the field of safety and risk management. The next step is the analysis of the design and functioning of chemical equipment and the features of the process. Further, an understanding of the data, their sources, measured parameters, and also operations and modes is required. Then data preparation, input, deletion, and formatting are carried out. The next step of the algorithm is the creation and simulation of an analytical model, for example, classification, clustering, regression, pattern recognition, etc., including using intelligent technologies. In conclusion, the model is evaluated, i.e., validation, evaluation, and testing with subsequent implementation.

At the initial stages of the algorithm, an important role is played by the types and classification of process safety data (Fig. 6.3). First of all, data can be divided into unstructured and structured data. Unstructured data (or unstructured information) is information that either does not have a predefined

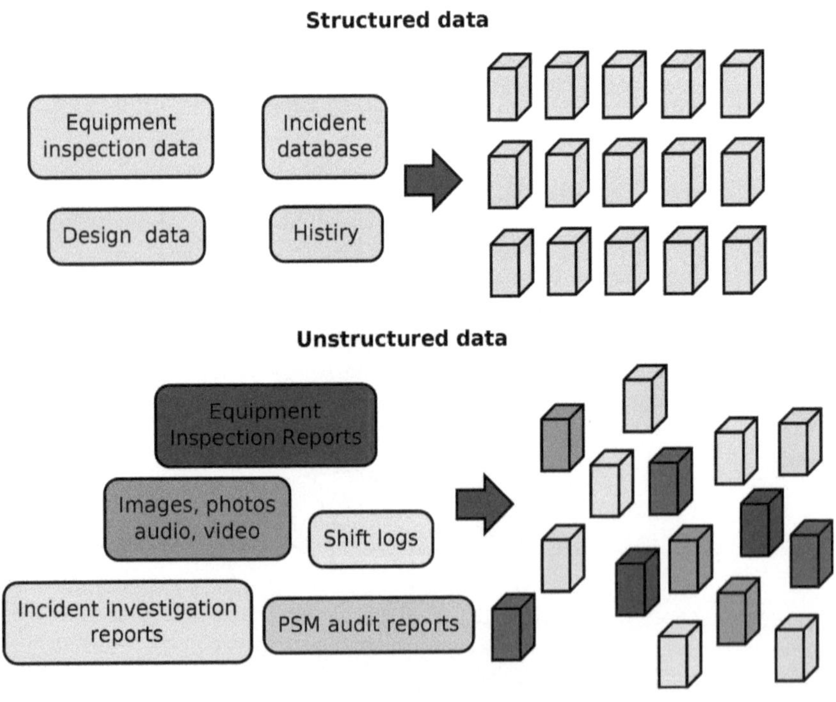

Fig. 6.3 Types and classification of process safety initial data.

data model or is not organized in a predefined way. Unstructured information, as a rule, contains a lot of text, but may also contain data such as dates, numbers, and facts. Structured data is a general term that refers to any organized data that conforms to a certain format.

Unstructured data includes mainly various types of reports, including incident investigation reports, process safety audit reports, equipment inspection reports, shift logs, shift handover verbal communications, look/listen/feel (LLF) observations, images, and audio and video files. The structured data includes incident databases, equipment verification data, data within corporate information systems (IT projects management—ITPM), design data such as a standard operating procedure (SOP), which is a set of step-by-step instructions compiled by an organization to help workers carry out complex routine operations, P&ID (piping and instrumentation diagram), and plot plans. Also, structured data includes historical data, for example, process parameters (operational data), alarms, event logs, equipment monitoring data (programmable logic controllers (PLC), distributed control system (DCS), and supervisory control and data acquisition systems (SCADA)), and some others.

Data is either static or dynamic. Static data is data that does not change. For example, the plant serial number or the commissioning dates. Dynamic data is subject to change. This is also called transactional data. Dynamic data may change after updating a related record. For example, distillation process parameters are constantly recorded and updated.

Examples of static data are process safety studies (PHA, HAZOP, incident investigation reports, emergency response plans), process safety management systems (audit reports, learning from incident communications, training records, safety culture assessments), and design data (P&IDs, SOPs, plot plans). Dynamic data refers to history (process parameters and production data), centralized maintenance management system (CMMS), laboratory information management system (LIMS), and operational data (work permits, mechanical integrity).

In conclusion, we note the advantages of using data analytics for process safety (Albalawi, Durand, & Christofides, 2017; Goel, Datta, & Mannan, 2017; Goel, Datta, & Sam Mannan, 2017; Goel, Pasman, Datta, & Mannan, 2019). Existing mechanisms include real-time risk assessment; development of an optimal equipment maintenance schedule; resource allocation; asset management; and installation of visual dashboards to increase the efficiency of dispatchers', engineers', and managers' work. In the future, the use of analytics will make possible incident forecasting, prevention of equipment/device failures, building a dynamic risk matrix, active analysis of text, audio, photo, and video, in addition to basic data analysis. For the successful implementation of these plans, it is necessary to solve the following problems: coordination of business requirements and requirements of standards; data quality and availability; data collection; and interdisciplinary integration. The following sections will cover the main methods and tools for data analytics, as well as examples of the use of these methods.

6.2 Machine learning

Machine learning is a class of artificial intelligence methods that is not a direct solution to problems. To implement such methods, mathematical statistics, numerical methods, measurement methods, probability theory, graph theory, and various methods of working with data in digital form are used (Al-Hadraawy, 2020; Carbonell, Michalski, & Mitchell, 1983; Gori, 2018; Theodoridis, 2015).

6.2.1 Machine learning basics

Machine learning implies that computers discover how they can perform tasks without being explicitly programmed for this. For simple tasks assigned to computers, you can program algorithms that tell the machine how to perform all the necessary steps to solve the task. Thus, from the computer side, training is not required. For more complex tasks, it can be difficult for a person to create the necessary algorithms manually. In practice, it may be more effective to help the machine develop its own algorithm, rather than human programmers, indicating each necessary step. The machine learning process, in general, can be divided into four interconnected components:

- data storage using surveillance, memory, and recall to provide a factual basis for further reasoning;
- abstraction involving translating stored data into a broader view and concept;
- generalization using abstract data to create knowledge and inferences that govern actions in the new concepts; and
- evaluation providing a feedback mechanism to measure the usefulness of acquired knowledge and to inform about possible improvements (Fig. 6.4).

These stages, presented in Fig. 6.4, characterize machine learning in the most general form and from a theoretical point of view. To solve practical

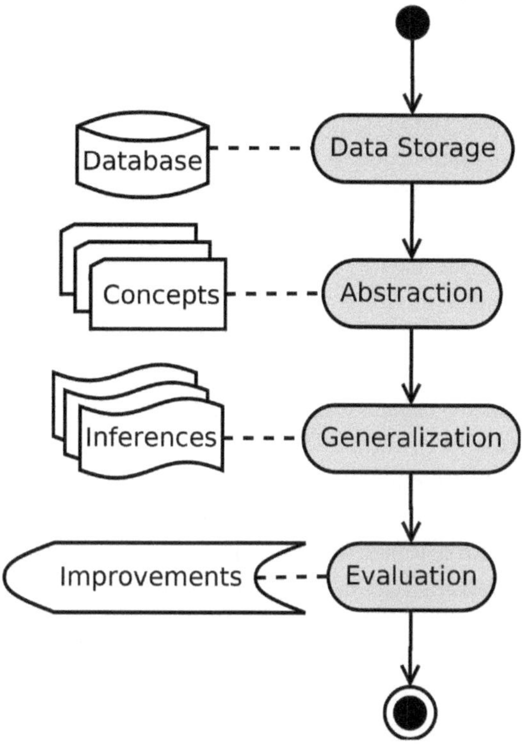

Fig. 6.4 The main stages of the machine learning process.

problems, the stages considered must be specified. Thus, any machine learning algorithm, some of which will be discussed later, can be implemented through the following steps (Bikmukhametov & Jäschke, 2020; Haesik, 2020; Lantz, 2019).

1. **Gathering data**. The data collection phase includes the collection of training materials that the algorithm will use to obtain effective knowledge. A dataset can be compiled from various sources, such as file, database, sensors, IoT, and many others. However, the collected data cannot be immediately used to perform the analysis process, since there can be a lot of missing data, extremely large values, disorganized text data, or data with noise. Inconsistent data might be collected due to human errors or duplication of data (errors with the name or values). In most cases, it will be advisable to combine the data in one source, such as a text file, spreadsheet, or database.
2. **Data exploration and preprocessing**. The quality of any machine learning project depends largely on the quality of its input. Thus, it is

important to learn more about data and its nuances through a practice called data mining. Data preprocessing is the process of cleaning up raw data. This means that data are collected in the real world and converted to a clean data set. In other words, whenever data are collected from different sources, it is collected in raw format, and this data is not suitable for analysis.

The following are some of the basic preprocessing methods that you can use to convert raw data:

- *Conversion of data*. Since we know that machine learning models can only process numerical functions, therefore, categorical and ordinal data must be somehow converted into numerical functions.
- *Ignoring missing values*. When identifying missing data in a set, we can delete a row or column of data depending on our needs. This method is known to be effective, but should not be used if there are many missing values in the dataset.
- *Filling in missing values*. When identifying missing data in a set, we can fill in the missing data manually. Most often, the average value of the missed value, the median, or the highest frequency is used.
- *Machine learning*. If there is some missing data, we can predict which data will be present in an empty position using existing data and approximation.
- *Outlier detection*. The data set under consideration may contain some erroneous data that differ sharply from other observations in the data set (for example, the temperature of catalytic cracking of crude oil at 4500°C was detected as erroneous data due to a typo by the operator, who added an extra "0").

3. **Model training**. By the time the data is ready for analysis, you are likely to have an idea of what you are able to extract from this data. The particular machine learning task selected will inform you of the selection of a suitable algorithm, and the algorithm will present data in the form of a model. You train the classifier using the "training dataset," adjust the settings using the "validation set," and then test the performance of your classifier on the invisible "test dataset."
4. **Assessment of the model**. Since each machine learning model provides a biased solution to the learning problem, it is important to evaluate how well the algorithm learns from its own experience. Depending on the type of model you are using, you can evaluate the accuracy of the model using a set of test data, or you may need to develop performance

indicators specific to the intended application. Model assessment is an integral part of the model development process. This helps to find the best model that represents the data, and how well the chosen model will work in the future.

5. **Improving the model**. If better performance is required, it becomes necessary to use more advanced strategies to improve model performance. Sometimes you may need to switch to another type of model entirely. To improve the model, we could adjust the model's hyperparameters and try to increase the accuracy, as well as look at the confusion matrix to try to increase the number of true positive and negative sides. You may need to supplement your data with additional data or perform additional preparatory work, as in the second stage of this process.

After these steps are completed, if the model seems to work well, it can be deployed to solve a specific practical problem. The successes and failures of the deployed model may even provide additional data for educating the next generation student. Depending on the circumstances, you can use your model to provide assessment data for forecasts of the level of process safety risk, equipment wear, and abnormal changes in process parameters, including in real time.

The practice of machine learning involves comparing the characteristics of the input data with the features of the available approaches to data analysis. Thus, before applying machine learning to solving real problems, it is important to understand the terminology that characterizes the input data sets.

The term *unit of observation* is used to describe the smallest object with measured properties of interest to the study. Typically, units of observation represent objects or things, transactions, measurements, or time periods. Examples include an enterprise, industrial plant, process temperature, and pressure. Sometimes units of observation are combined into units such as enterprise-years, which indicate cases where the same enterprise has been tracked for several years; each enterprise-year consists of data on the enterprise for 1 year. A unit of observation refers to the category, type, or classification to which everyone or everything belongs, and not to specific people or objects.

The unit of observation is associated, but not identical, with the *unit of analysis*, which is the smallest unit from which the conclusion is drawn. In some cases, the observed and analyzed units are identical, but in others, they may not coincide (Sedgwick, 2014). A unit is simply what or who needs to be described or analyzed. For example, the testing data on the process safety training for employees of a manufacturing enterprise can be recorded for individual employees (in this case, the unit of measurement is the same as the unit of observation). Alternatively, you can group employees by region or state and compare Texas process workers with California employees, thereby creating a new unit of analysis (for example, employee groups). Thus, data obtained from individuals can be used to analyze trends in different regions or countries.

Datasets in which observation units and their properties are stored can be represented as datasets consisting of two components. The first component is *examples*, which are instances of the unit of observation for which properties have been recorded. The second component is *features*, or recorded properties or attributes of examples, that may be useful for machine learning.

Consider the features and examples used in machine learning to solve real problems. In order to create a training algorithm for identifying critical situations in the catalytic cracking process, a critical situation can be used as a unit of observation, using examples of specific critical situations that arose at the facility, and features can consist of specific parameters characterizing the cracking process, such as temperature, pressure, and input velocity flow. For an algorithm for detecting hidden equipment defects, the unit of observation can be equipment; examples may include random sampling of failed equipment, and features may include types of failures, as well as equipment characteristics such as the intensity and type of load, maintenance personnel, and operating conditions.

Typically, examples and features do not need to be collected in any particular form. However, traditionally they are assembled in a matrix format, which means that each example has exactly the same set of features. Fig. 6.5 shows a data set in matrix format.

In matrix data, each row in a spreadsheet is an example, and each column is a feature. The rows in Fig. 6.5 indicate examples of accidents, while columns record various characteristics of each accident, such as plant model, accident type, cause, and damage. Data in matrix format is by far the most common form used in machine learning algorithms. In rare special cases, other forms are sometimes used.

Features also come in many forms. If a feature is a characteristic measured in numbers, it is called *numerical*, whereas if the feature is an attribute consisting of a set of categories, the feature is called *categorical* or *nominal*. A special case of categorical variables is called *ordinal*, which denotes a nominal variable with categories falling into an ordered list. Examples of ordinal variables are qualitative risk assessments such as low, medium, and high; or

Date	Plant model	Accident type	Cause	Damage
2005	XY234	Explosion	Pump failure	Strong
2010	YZ123	Fire	Pressure sensor failure	Medium
2017	XZ235	Leak	Corrosion	Minor
2001	XY234	Fire	Corrosion	Minor
2012	YZ123	Explosion	Pump failure	Medium
2015	YZ235	Leak	Corrosion	Minor
2009	XY234	Explosion	Pressure sensor failure	Medium
2014	YZ235	Fire	Valve failure	Strong

Examples

Features

Fig. 6.5 Input data presented in matrix format.

measuring the frequency of incidents on a scale of very rarely, rarely, infrequently, often, and very often. It is important to consider what the features are, since the type and number of features in the data set will help determine the appropriate machine learning algorithm for a particular problem.

6.2.2 Models and tasks of machine learning

Machine learning algorithms can be divided into categories according to their purpose. Understanding the categories of learning algorithms and the ability to choose the right category is an important step in using data to manage process safety risks.

The *predictive model* is used for tasks in which it is necessary to build a forecast of one value using other values in the data set. The learning algorithm attempts to detect and model the relationship between the *target feature* (predicted feature) and other features. Despite the usual use of the word "prediction" forecasting, prognostic models need not anticipate future events. This may include past events or real-time process control.

For example, a predictive model can be used to predict past events, such as the point at which industrial equipment failure begins to develop using the current values of the process parameters. Predictive models can also be used in real time to manage the energy consumption of production halls to avoid network congestion. Since predictive models are given clear instructions about what they need to learn and how they should be learned, this learning process of the predictive model is called *supervised learning*. The supervision does not relate to human participation, but rather to the fact that the target values give the learning system the opportunity to find out how well it has learned the desired task. In other words, having at its disposal a certain set of input data, the supervised learning algorithm tries to optimize the function (model) in order to find a combination of feature values that lead to the target output. Fig. 6.6 presents the correspondence of models and tasks of machine learning (Lantz, 2019).

Quite often, the forecasting problem is used in the framework of controlled machine learning, in which it is necessary to determine which

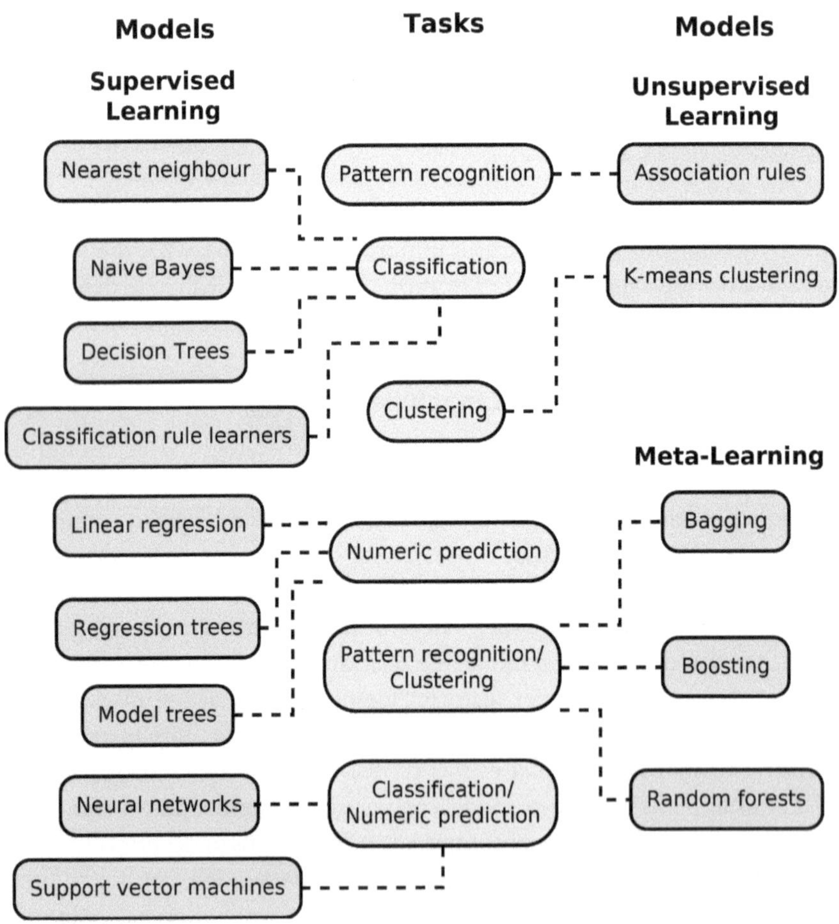

Fig. 6.6 Models and tasks of machine learning.

category the example belongs to. Such a task is called *classification*. Thus, classification is one of the sections of machine learning dedicated to solving the following problem. There are many objects (situations) divided in some way into classes. A finite set of objects is given for which it is known which classes they belong to. This set is called the training set. The class affiliation of the remaining objects is not known. It is necessary to construct an algorithm capable of classifying an arbitrary object (example) from the original set. To classify an object means to indicate the number (or class name) to which the object belongs. In other words, the task of classification is to obtain a categorical answer based on a set of features. It has a finite number of answers (usually in the yes or no format): is there a fire in the photo, is the image a human face. For example, you can predict the following:

- Optical text recognition: from the scanned image of the text to determine the chain of characters that form it.
- Synthesis of chemical compounds: according to the parameters of chemical elements to predict the properties of the resulting compound.
- Situation analysis: is a critical situation possible with a given combination of chemical process parameters?

In classification, the target feature to be predicted is a categorical feature. This feature is known as a *class*, and is divided into categories called *levels*. Levels may or may not be ordinal. A class can have two or more levels. The following classification types can also be distinguished:

- Two-class classification. The most technically simple case, which serves as the basis for solving more complex problems.
- Multiclass classification. When the number of classes can reach many thousands (for example, when recognizing hieroglyphs or continuous speech), the task of classification becomes significantly more difficult.
- Disjoint classes.
- Intersecting classes. An object can belong to several classes at the same time.
- Fuzzy classes. It is required to determine the degree of belonging of an object to each of the classes; usually this is a real number from 0 to 1.

Since classification is very widely used in machine learning, there are many types of classification algorithms, with strengths and weaknesses suitable for different types of input data.

Supervised learning can also be used to predict numerical data, such as risk, laboratory metrics, test results, or number of items or events.

To predict such numerical values, linear regression models for input data are considered the most common form of *numeric prediction*. Although regression models are not the only type of numeric model, they are by far the most widely used. Regression methods are widely used for forecasting, since they accurately quantify the relationship between the input data and the target, including both the magnitude and the uncertainty of the relationship.

Since numbers can easily be converted into categories (for example, employees from 45 to 60 years old) and categories into numbers (for example, assign 1 to all critical situations, 0 to all regular situations), the boundary between classification models and numeric predictive models can be very blurred.

Below is a brief summary of the well-known method of stepwise additive modeling of numeric prediction (Kotu & Deshpande, 2019; Witten, Frank, Hall, & Pal, 2017). First, a standard regression model is created, for example, a regression tree; the errors that it demonstrates in the training data—the differences between the predicted and observed values—are called *residuals*. Then you need to correct these errors by examining the second model—perhaps another regression tree—that tries to predict the observed residuals. To do this, simply replace the original class values with their remnants before studying the second model. Adding forecasts made by the second model to the forecasts of the first automatically reduces the errors in the training data. Usually, some residuals still remain because the second model is not perfect. Thus, it is necessary to use a third model, which learns to predict residual balances and so on.

If individual models minimize the squared error of prediction, as linear regression models usually do, this algorithm minimizes the squared error of the ensemble as a whole. In practice, this also works well when a basic learner uses heuristic approximation instead, such as the regression and model tree learners.

Descriptive models relate to the recording and analysis of statistical data to enhance the capabilities of business intelligence. Managers and safety engineers get a description and the most informative analysis of the results and consequences of past actions and decisions. This process is currently characteristic of most large manufacturing enterprises, for example, an analysis of process safety management procedures to determine their efficiency. Within the framework of descriptive models, data is generalized in new nonclassical ways. Unlike predictive models, which predict the behavior of the object under study, in a descriptive model, all the features are equally important. In this case, the goal of learning is missing, so the learning process of a descriptive model is called *unsupervised learning*.

For example, *pattern recognition* is the automated recognition of patterns and regularities in data. It has applications in statistical data analysis, signal processing, image analysis, information retrieval, bioinformatics, data compression, computer graphics, and machine learning. Pattern recognition has its origins in statistics and engineering; some modern approaches to pattern recognition include the use of machine learning, due to the increased availability of big data and a new abundance of processing power. However, these activities can be viewed as two facets of the same field of application, and together they have undergone substantial development over the past few decades. A pattern recognition is based on the automatic discovery of regularities in given data through applying computer algorithms and classifying the data into different categories with help of the discovered regularities (Golden, 2015).

Pattern recognition is generally categorized according to the type of learning procedure used to generate the output value. Supervised learning assumes that a set of training data (the training set) has been provided, consisting of a set of instances that have been properly labeled by hand with the correct output. A learning procedure then generates a model that attempts to meet two sometimes conflicting objectives: perform as well as possible on the training data, and generalize as well as possible to new data (usually, this means being as simple as possible, for a technical definition of "simple," in accordance with Occam's Razor, discussed below). Unsupervised learning, on the other hand, assumes training data that has not been hand-labeled, and attempts to find inherent patterns in the data that can then be used to determine the correct output value for new data instances (Carvalko & Preston, 1972). A combination of the two that has recently been explored is semisupervised learning, which uses a combination of labeled and unlabeled data (typically a small set of labeled data combined with a large amount of unlabeled data). Note that in cases of unsupervised learning, there may be no training data at all to speak of; in other words, the data to be labeled is the training data (Uttal, 2002).

Cluster analysis is the task of grouping many objects into subsets (clusters) so that objects from one cluster are more similar to each other than objects from other clusters by some criterion. The clustering problem belongs to the class of teaching tasks without a teacher (Pimentel & de Carvalho, 2020).

Cluster analysis performs the following main tasks: typology or classification development; a study of useful conceptual schemes for grouping objects; generation of hypotheses based on data research; and hypothesis testing or research to determine whether the types (groups) identified in one way or another are actually present in the available data.

Clustering is used in image segmentation to determine boundaries and object recognition. Cluster analysis can be applied for selection of recommendations for the user based on the known preferences of other users from this cluster.

You can find a description of two fundamental requirements for data—uniformity and completeness. Homogeneity requires that all clustered entities be of the same nature, described by a similar set of characteristics.

Finally, there is one more specific class of machine learning algorithms. *Meta learning* is a subfield of machine learning where automatic learning algorithms are applied on metadata about machine learning experiments (Schaul & Schmidhuber, 2010). As of 2017, the term had not found a standard interpretation, however the main goal is to use such metadata to understand how automatic learning can become flexible in solving learning problems, hence improving the performance of existing learning algorithms or learning (inducing) the learning algorithm itself, hence the alternative term "learning to learn" (Lemke, Budka, & Gabrys, 2013).

Flexibility is important because each learning algorithm is based on a set of assumptions about the data—its inductive bias (Schaul & Schmidhuber, 2010). This means that it will only learn well if the bias matches the learning problem. A learning algorithm may perform very well in one domain but not in another. This poses strong restrictions on the use of machine learning or data mining techniques, since the relationship between the learning problem (often some kind of database) and the effectiveness of different learning algorithms is not yet understood.

Lastly, a class of machine learning algorithms known as metalearners is not tied to a specific learning task, but is rather focused on learning how to learn more effectively. A metalearning algorithm uses the result of some learnings to inform additional learning. This can be beneficial for very challenging problems or when a predictive algorithm's performance needs to be as accurate as possible.

Thus, faced with the real task of data analysis, it is first necessary to determine which of the considered main problem types it refers to (Fig. 6.6). The researcher must find out if this will be a classification, numeric prediction, clustering, or pattern recognition. Then, in accordance with the selected type of task, the researcher must determine the model of machine learning. So, if the task facing the analyst is a clustering task, then for its solution, the *k*-means model or one of the metalearning algorithms can be selected. To solve classification problems, decision trees, the nearest neighbor, or neural networks can be used. If the problem being solved is of the numeric prediction type, then linear regression, regression trees, or model trees can be used as a possible model.

All the considered methods of machine learning have a different level of complexity. It can be assumed that the most effective strategy is to use the most complex models of machine learning, for example, from neural networks, which act as a universal approximator. However, there are several reasons why this should not always be done.

Firstly, simple methods, such as decision trees, the method of nearest neighbors or a random forest, are more quickly tuned to the data, which allows you to quickly get the first forecasting results, understand the specifics of the problem, and evaluate for which objects the forecasting quality is high and for which it is low. Thus, it is possible to more accurately focus the attention of analysts on the problematic aspects of the task, namely, on the types of objects for which the algorithm works the worst, and not try to configure the algorithm as a whole.

Secondly, simple machine learning methods are easier to interpret. Using a simpler and more intuitive model, it is easier to understand the logic of their work and explain why one or another forecast was obtained for each set of attributes (features). Interpretability allows you to better understand the internal dependencies in the data, i.e., what types of features most affect the forecast, and what type of this influence (linear, nonlinear, independent, or in combination with other features). This allows you to select the most significant features for prediction, and then use them in more complex models.

In some cases, even though a complex model can provide greater accuracy, the researcher is limited only to simple interpretable models. This situation occurs when the forecast is associated with the adoption of decisions for which the cost of error is very high. This is especially true when it comes to decision-making in the field of process safety, when building a forecast for the development of a critical situation or the effectiveness of proposed risk management measures. In such cases, one cannot rely on one analyst with one uninterpreted machine learning algorithm, and the value of the algorithm begins to be determined not only by the accuracy of the forecast, but also by the ability to understand and explain the logic of its work in order to make adjustments on the part of subject matter experts (in this case, process safety specialists).

The results of most machine learning algorithms (for example, the method of nearest neighbors, *k*-means) depend on the scale of the features, and if you want to guarantee the equal contribution of the features to the prediction, you need to normalize them by subtracting the average and dividing by the standard deviation. Otherwise, changes in signs with a small spread may go unnoticed in the context of changes in features with a wide spread of accepted values. For features that have a highly nonuniform distribution, a nonlinear monotonic transformation may be required. For example, when analyzing texts, the number of words used is most often equal to zero or a small amount, but for some documents certain words can occur very often. In order to smooth out the uneven distribution of the number of words, we can consider the logarithm of the initial value, which reduces the values of the attribute that are too large. Most methods do not work directly with discrete characters, and they need to be converted to binary using specific encoding.

Thus, data preparation for the selected algorithm is the next important stage in the implementation of the analytical project. At the same time, it is important not only to understand the existing data and select those related to the problem being solved, but also to convert them to a format suitable for a particular machine learning algorithm.

In the following sections, we shall consider some particular tasks and models of machine learning. Once you get an idea of basic algorithms, you will be able to choose and work with other models not discussed in the book.

6.3 Basic data analytics

The relatively simple data analysis methods discussed in this section are mainly based on the classical postulates of probability theory and mathematical statistics. Among these analytical methods are, first of all, clustering, classification, and numerical prediction.

6.3.1 Clustering

Clustering is a method often used for exploratory analysis of data. In clustering, there are no predictions made. Rather, clustering methods find the similarities between objects according to the object features, and group the similar objects into clusters (Celebi, Kingravi, & Vela, 2013; Kriegel, Schubert, & Zimek, 2017; Pelleg & Moore, 1999). Clustering or cluster analysis can be defined as the use of unsupervised methods of grouping

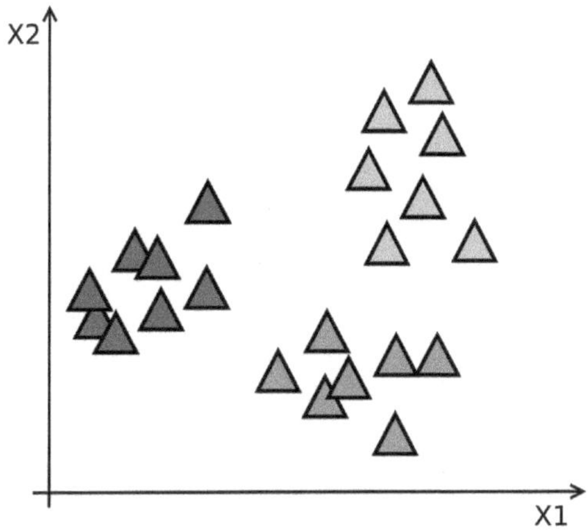

Fig. 6.7 The result of cluster analysis in two-dimensional space is shown as colored triangles in three clusters.

similar objects (Fig. 6.7). In the figure, triangles belonging to one cluster are colored with one color. So, here we can see three clusters. In machine learning, the term "unsupervised" is related to the problem of finding a hidden structure in unlabeled data. When clustering models are said to be unsupervised, it means that the analyst does not predetermine the labels for application to the clusters.

Clustering has certain specific goals. First of all, this is understanding data by identifying a cluster structure. Dividing the sample into groups of similar objects allows us to simplify further data processing and decision making by applying a different analysis method to each cluster (divide and conquer strategy). Second is data compression. If the original sample is excessively large, then you can reduce it by leaving out one of the most typical representatives from each cluster. Finally, there is novelty detection. Atypical objects are selected that cannot be attached to any of the clusters.

In the first case, the intention is to make the number of clusters smaller. In the second case, it is more important to ensure a high (or fixed) degree of similarity of objects within each cluster, and there can be as many clusters as possible. In the third case, individual objects that do not fit into any of the clusters are of the greatest interest. In all these cases, hierarchical clustering can be applied, when large clusters are split up into smaller ones, which in turn are split up even smaller, etc. Such tasks are called taxonomy problems.

The result of taxonomy is a tree-like hierarchical structure. Moreover, each object is characterized by listing all the clusters to which it belongs, usually from large to small. Visually, taxonomy is presented in a graph called a dendrogram. As part of clustering, the following types of input data are distinguished:

- Characteristic description of objects. Each object is described by a set of its characteristics, called features. Features can be numeric or nonnumeric.
- A matrix of distances between objects. Each object is described by the distances to all other objects of the training set.

The distance matrix can be calculated from the matrix of feature descriptions of objects in an infinite number of ways, depending on how you enter the distance function (metric) between feature descriptions. The Euclidean metric is often used, but in most cases this choice is heuristic and is due only to considerations of convenience.

The inverse problem is restoration of feature descriptions from the matrix of pair-wise distances between objects and, in general, has no solution, and the approximate solution is not unique and may have a significant error. This problem is solved by multidimensional scaling methods.

Thus, the statement of the clustering problem by the distance matrix is more general. On the other hand, when there are indicative descriptions, it is often possible to build more efficient clustering methods.

The data structure describes the objects of interest and determines how best to group the objects. For example, based on normalized process safety risk levels of different enterprises, it is easy to divide enterprises into three groups depending on arbitrary values. Enterprises can be divided into three groups as follows:

- risk level less than 0.2;
- risk level between 0.2 and 0.899; and
- risk level 0.9 or more.

There are many models or algorithms for clustering; some of them were mentioned in Section 6.2.2. One of the popular clustering algorithms is *k*-means. Let's describe it in brief.

Clustering of *k*-means is a vector quantization method, initially based on signal processing, which is aimed at dividing *n* observations into *k* clusters in which each observation belongs to a cluster with the closest *mean* (cluster centers or clusters centroid), which serves as a prototype cluster. This leads to the splitting of the data space into Voronoi cells. It is popular for cluster analysis in data mining. Clustering *k*-means minimizes variances within the cluster (squared Euclidean distances), but not regular Euclidean distances, which would be Weber's more difficult task: the mean optimizes square errors, while only the geometric median minimizes Euclidean distances. For example, the best Euclidean solutions can be found using *k*-medians and *k*-medoids.

The problem is computationally complex (*NP*-hard); however, effective heuristic algorithms quickly converge to a local optimum. They are usually similar to the expectation maximization algorithm for mixtures of Gaussian distributions using an iterative refinement approach used for both *k*-means and for modeling a Gaussian mixture. They both use cluster centers to model data; however, *k*-means clustering tends to find clusters of comparable spatial extent, while the expectation maximization mechanism allows clusters to have different shapes.

The *k*-means algorithm has the following mathematical description. Given a set of observations $(\mathbf{x}_1, \mathbf{x}_2, \ldots, \mathbf{x}_n)$, where each observation is a *d*-dimensional real vector, *k*-means clustering aims to partition the *n* observations into $k(\leq n)$ sets $\mathbf{S}=\{S_1, S_2, \ldots, S_k\}$ so as to minimize the within-cluster sum of squares (WCSS) (i.e., variance). Formally, the objective is to find minimum of function:

$$V = \underset{S}{\arg\min} \sum_{i=1}^{k} \sum_{x \in S_i} \|x - \mu_i\|^2,$$

where μ_i is the mean of points in S_i. Because the total variance is constant, this is equivalent to maximizing the sum of squared deviations between points in different clusters (between-cluster sum of squares, BCSS), which follows from the law of total variance (Kriegel et al., 2017).

The algorithm described above finds the clusters and dataset labels for a specific preselected *k*. To find the number of clusters in the data, the user must run the *k*-means clustering algorithm for the range of *k* values and compare the results. Here are some of *k*-means' possible usages (Trevino, 2016):

- Sort measurement sensors, i.e., identify activity types in motion sensors; group images; separate audio; identify groups in safety and health monitoring.
- Detection of bots or anomalies: separate valid activity groups from bots; group valid actions to clear outlier detection.

In addition, *k*-means can help in observing if the monitored data point switches between groups over time and can be used to detect significant changes in the data. However, the method has the following disadvantages:

- The global minimum of the total quadratic deviation *V* is not guaranteed to be achieved, but only one of the local minima.

- The result depends on the choice of the initial cluster centers; their optimal choice is unknown.
- The number of clusters must be known in advance.

6.3.2 Classification

Together with analytical techniques, such as clustering, classification is another basic learning method that is used to solve data mining problems. When classification learning, the classifier is presented along with a set of examples that are already classified. The main task solved by the classifiers is to assign class labels to new observations or examples. Most classification methods are supervised and start with a training set of prelabeled cases (examples or observations) to find out how likely the attributes of these cases are to help classify future unlabeled cases.

In the framework of process safety management and risk assessment, classification techniques are useful to help decision makers decide between options that involve multiple risks and where compromises have to be made. Methods help provide a logical basis for the rationale for a decision (MacQueen, 1967).

Consider the mathematical formulation of the classification problem. Let X be a set of descriptions of objects, and Y be a set of class numbers (or names). There is an unknown target dependence—a mapping y^*: $X \to Y$—of which values are known only as the objects of the final training sample $X^m = \{(x_1, y_1), \ldots, (x_m, y_m)\}$. It is necessary to build an algorithm a: $X \to Y$ capable of classifying an arbitrary object x from the set X.

More general is the probabilistic statement of the problem. It is assumed that the set of "object, class" pairs $X^{\times}$ Y is a probability space with an unknown probability measure P. There is a finite training sample $X^m = \{(x_1, y_1), \ldots, (x_m, y_m)\}$ of observations generated according to the probability measure P. It is necessary to construct an algorithm a:$X \to Y$ capable of classifying an arbitrary object x from the set X.

A feature is a mapping, where D_f is the set of admissible values of the feature. If the signs $f_1, \ldots, f_n$ are given, then the vector $\mathbf{x} = (f_1(x), \ldots, f_n(x))$ is called the feature description of the object x from the set X. Feature descriptions can be identified with the objects themselves. Moreover, the set $X = D_{f1} \times \cdots \times D_{fn}$ is called feature space. Depending on the set of D_f, the features are divided into the following types:

- binary feature: $D_f = \{0,1\}$;
- nominal feature: D_f is finite set;

- ordinal feature: D_f is a finite ordered set; and
- quantitative feature: D_f is a set of real numbers.

Often there are applied problems with heterogeneous features, not all methods are suitable for solving them. This section is dedicated to the basic more common classification methods, i.e., *decision trees* and *naive Bayes*.

As mentioned above, the ability to model unknown complex interactions between variables has made machine learning a common tool in computer science and computational methods. Among other things, decision trees have been used locally for several successful applications that achieved a high level of performance. In addition, decision trees provide an opportunity to check decision rules and examine the relevance of each variable, as well as the relationship between them.

The decision tree models the possible paths that follow from the initial decision that must be made (for example, whether to continue project A or project B). As two hypothetical projects are implemented, a series of events may occur, and various predictable decisions will need to be made. They are presented in a tree format similar to an event tree (Fig. 6.8). The probability of events can be estimated along with the expected value or usefulness of the final result of each path (Fratello & Tagliaferri, 2018; Steven, Brown, & Myles, 2019).

The decision tree can be used to build structure and make consistent decisions. This is especially useful if the complexity of the problem increases. This approach allows managers and safety professionals to quantify the possible outcomes of decisions and, therefore, helps decision-makers to choose

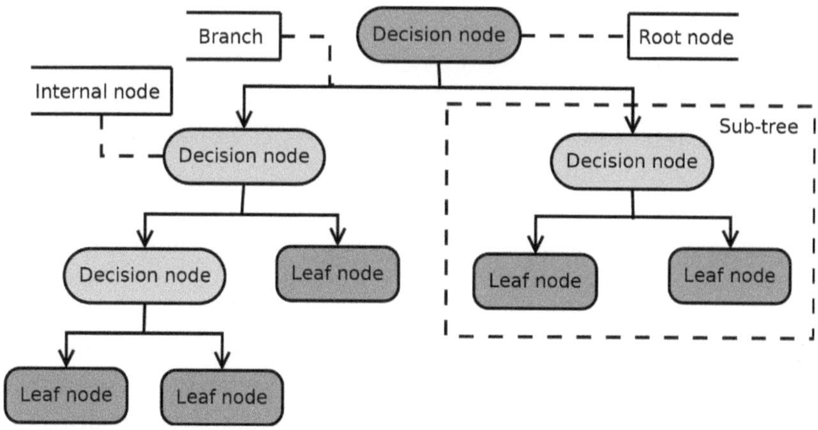

Fig. 6.8 The main components of a decision tree.

the best course of action when the results are uncertain. Visualization of the decision tree in the form of a graph can also help in substantiating the reasons for the decisions made. Information about the path to the best solution is the one that gives the best expected value, calculated as the product of all conditional probabilities along the path and the value of the result. The method often uses subjective estimates of event probabilities, and can be used for short-, medium-, and long-term issues at the operational or strategic level.

To develop a decision tree, a plan of possible actions is required with decision points, and information about the possible results of decisions and random events that may affect decisions. Experience is required to configure the tree properly, especially in difficult situations. Depending on the design of the tree, quantitative data or sufficient information is necessary to justify the opinions of experts on probabilities. The outputs of the decision tree include:

- graphical representation of the solution to the problem;
- calculation of the expected value for each possible path; and
- a priority list of possible outcomes based on the expected value, or a recommended path to follow.

The strengths of the method include the following:

- It provides a clear graphical representation of problem solving details;
- It allows you to calculate the best way through the situation and the expected result;
- It develops clear thinking and planning; and
- A tree design exercise can lead to a better understanding of the problem.

However, the method has the following limitations:

- Large decision trees may become too complex for easy discussion;
- There may be a tendency to simplify the situation so that it can be represented in the form of a tree diagram;
- It relies on historical data that may not apply to the simulated solution; and
- The method simplifies the problem-solving approach, which eliminates extreme values.

The next classification method, *naive Bayes*, assumes availability of good understanding of events consequences and their likelihood (or probabilities).

The probability of an event or a specific consequence can be estimated by extrapolating historical data (provided that sufficient relevant historical data is available so that the analysis is statistically reliable). This is especially true for zero cases where it cannot be assumed that, since an event or consequence did not happen in the past, it will not happen in the near future.

Another tool is modeling methods for generating, for example, the likelihood of breakdowns of equipment and structures due to aging and other degradation processes. You can ask experts to give their opinion on the likelihood and consequences, given the relevant information and historical data.

Consequences and likelihood can be combined to give a level of risk. This can be used to assess the significance of the risk by comparing the risk level with the acceptance criterion or in order to rank the risks in order.

A naive Bayes classifier is a simple probabilistic classifier based on the application of Bayes' theorem with strong (naive) independence assumptions between the features (Domingos & Pazzani, 1997; Maron, 1961). Naive Bayes models assign class labels to problem instances, represented as vectors of feature values, where the class labels are drawn from some finite set. There is no single algorithm for training such classifiers, but a family of algorithms based on a common principle: all naive Bayes classifiers assume that the value of a particular feature is independent of the value of any other feature, given the class variable. For example, a fruit may be considered to be an orange if it is orange, round, and about 9 cm in diameter. A naive Bayes classifier considers each of these features to contribute independently to the probability that this fruit is an orange, regardless of any possible correlations between the color, roundness, and diameter features.

Depending on the exact nature of the probabilistic model, naive Bayes classifiers can be trained very effectively. In many practical applications, the maximum likelihood method is used to estimate parameters for naive Bayesian models; in other words, you can work with a naive Bayesian model without believing in Bayesian probability and without using Bayesian methods.

Despite the naive look and, undoubtedly, very simplified conditions, naive Bayes classifiers often work much better in many difficult practical situations (Berrar, 2018; Zhang & Sakhanenko, 2019). The advantage of the naive Bayes classifier is the small amount of data necessary for training, parameter estimation, and classification.

Let's consider the probabilistic model of the naive Bayes method. Abstractly, naive Bayes is a conditional probability model: given a problem instance to be classified, represented by a vector $\mathbf{x} = (C_k | x_1, \ldots, x_n)$ representing some n features (independent variables). It assigns to this instance probabilities $p(C_k | x_1, \ldots, x_n)$ for each of K possible outcomes of classes C_k. If the number of features n is large or if a feature can take on a large number of values, then basing such a model on probability tables is infeasible. We therefore reformulate the model to make it more tractable. Using Bayes' theorem, the conditional probability can be decomposed as:

$$p(C_k|\,\mathbf{x}) = \frac{p(C_k)p(\mathbf{x}|\,C_k)}{p(\mathbf{x})}.$$

In practice, there is interest only in the numerator of that fraction, because the denominator does not depend on C and the values of the features x_i are given, so that the denominator is effectively constant. The numerator is equivalent to the joint probability model $p(C_k,\ x_1,\ \ldots,\ x_n)$, which can be rewritten as follows, using the chain rule for repeated applications of the definition of conditional probability:

$$p(C_k, x_1, \ldots, x_n) = p(x_1, \ldots, x_n, C_k) = p(x_1|\,x_2, \ldots, x_n, C_k)p(x_2, \ldots, x_n, C_k)$$
$$= L = p(x_1|\,x_2, \ldots, x_n, C_k)p(x_2|\,x_3, \ldots, x_n, C_k)\ldots p(x_{n-1}|\,x_n, C_k)p(x_n, C_k)p(C_k).$$

Now we introduce the "naive" assumptions about conditional independence: suppose that all the features in **x** are mutually independent and depend on the category C_k. According to this assumption $p(x_i\,|\,x_{i+1},\ldots, x_n, C_k) = p(x_i\,|\,C_k)$.

Thus, the joint model can be expressed as $p(C_k, x_1, \ldots, x_n) = p(C_k)\, p(x_1\,|\,C_k)p(x_2\,|\,C_k)\ldots p(x_n\,|\,C_k) = p(C_k)\prod_{i=1}^{n} p(x_i\,|\,C_k)$.

This means that under the above independence assumptions, the conditional distribution over the class variable C is:

$$p(C_k|\,x_1, \ldots, x_n) = \frac{1}{z}p(C_k)\prod_{i=1}^{n} p(x_i|\,C_k),$$

where the evidence $Z = p(\mathbf{x}) = \sum_k p(C_k)p(\mathbf{x}\,|\,C_k)$ is a scaling factor dependent only on $x_1,\ \ldots,\ x_n$, that is a constant if the values of the feature variables are known.

The naive Bayes classifier combines this probability model with a decision rule. One common rule is to pick the hypothesis that is most probable; this is known as the maximum a posteriori (MAP) decision rule. The corresponding classifier, a Bayes classifier, is the function that assigns a class label $\hat{y} = C_k$ for some k as follows:

$$\hat{y} = \underset{k\in\{1\ldots K\}}{\arg\max}\ p(C_k)\prod_{i=1}^{n} p(x_i|\,C_k).$$

A class's prior may be calculated by assuming equiprobable classes (i.e., priors = 1/<number of classes>), or by calculating an estimate for the class probability from the training set (i.e., <prior for a given class> = ≤number of samples in the class>/<total number of samples>). To estimate the parameters for a feature's distribution, one must assume a distribution or generate nonparametric models for the features from the training set (Caruana & Niculescu-Mizil, 2006; John & Langley, 1995).

The assumptions on distributions of features are called the "event model" of the naive Bayes classifier. For discrete features like the ones encountered in document classification (include spam filtering), multinomial and Bernoulli distributions are popular.

In spite the fact that far-reaching assumptions about independence are often inaccurate, the naive Bayes classifier has a number of properties that make it rather useful in practice. In particular, decoupling the distributions of conditional class attributes means that each distribution can be independently evaluated as a one-dimensional distribution. This helps alleviate the problems associated with the curse of dimensionality, for example, the need for data sets that exponentially scale depending on the number of features.

6.3.3 Numeric prediction

Predictive modeling is primarily concerned with minimizing model errors or, in other words, forecasting as accurately as possible. As mentioned earlier, machine learning borrows algorithms from various fields, including statistics, and uses them for this purpose. *Linear regression* is one of the most well-known and understandable algorithms in statistics and machine learning (Angelini, 2018; Ganesh, 2010). The general purpose of *multiple regression* is to analyze the relationship between several independent variables (also called *regressors* or *predictors*) and the dependent variable. Linear regression can be represented as an equation that describes a line that most accurately shows the relationship between the input variables x and the output variables y. To compile this equation, it is necessary to find certain coefficients b for the input variables, for example, in the case of the presence of one independent variable (the so-called *simple regression*) $\hat{y}_i = b_0 + b_1 x$ (Fig. 6.9).

The values $\hat{y}_i$ (points on the regression line) are called *estimates* of the output variables y_i, since they are estimated (calculated) by a linear model and differ from the real values of y_i by the deviations r_i. To evaluate the regression model, various linear algebra methods or the least squares method are used. Next, we will consider in more detail the method of constructing a regression model and its estimation from the point of view of suitability for obtaining numerical forecasts.

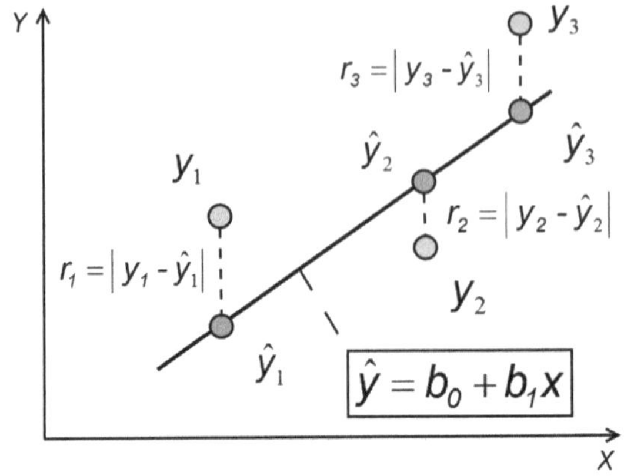

Fig. 6.9 Linear regression model.

In general, multiple regression allows the researcher to ask a question (and probably get an answer) about "what is the best predictor for…" For example, a process safety researcher might want to know which factors are the best predictors of equipment failure or a critical situation. Note that the term "multiple" indicates the presence of several predictors or regressors that are used in the model.

So, regression is a representation of a data set using some function $y(x)$, i.e., the task of regression is to select the parameters for this function in such a way that it represents a "cloud" of starting points defined by the vectors X and Y with a minimum mean square error. The fit of a regression model can be visualized by plotting the predicted data values against the actual values.

In other words, the general computational problem that needs to be solved in the analysis by the multiple regression method is to fit a straight line to a certain set of points (Fig. 6.10). In the simplest case, when there is one dependent and one independent variable, this can be seen in the *scattering diagram*. In the figure, you can see the values of the coefficients of the regression equation calculated on the basis of the *least squares* method. The regression line is constructed so as to minimize the squares of deviations of this line from the observed points. Therefore, this general procedure is sometimes referred to as least squares estimation.

The initial data and the results of the regression analysis for the presented example for a given confidence interval of 95% are shown in Fig. 6.11. The main indicators for assessing the prognostic effectiveness of the regression model are the t-statistics and the Fisher test F. These indicators will be discussed later.

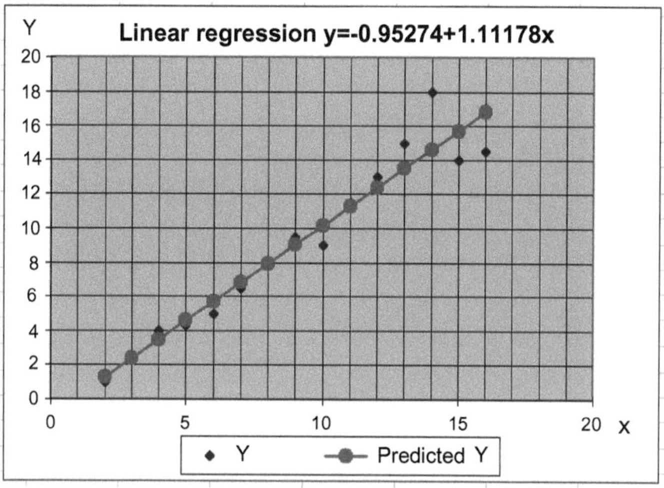

Fig. 6.10 An example of a graphical representation of linear regression.

Observations

X	2	3	4	5	6	7	8	9	10	11	12	13	14	15	16
Y	1	2,5	4	4,3	5	6,5	8	9,5	9	11,5	13	15	18	14	14,5

Regression statistics	
Multiple correlation R	0,965470904
R-squared	0,932134066
Normalized R-squared	0,926913609
Standard error	1,392243361
Observations	15

Analysis of variance

	df	*SS*	*MS*	*F*	*Significance F*
Regression	1	346,0988929	346,0989	178,5541	5,6956E-09
Residue	13	25,19844048	1,938342		
Total	14	371,2973333			

	Coefficients	*Standard error*	*t-statistic*	*P-value*	*Lower 95%*	*Top 95%*
Y intersection	-0,952738095	0,830636683	-1,147	0,272055	-2,7472195	0,841743
X	1,111785714	0,083202455	13,36241	5,7E-09	0,93203774	1,291534

Fig. 6.11 The initial data and the results of the regression analysis of a numerical example.

So, a straight line on the plane (in the space of two dimensions) is given by the equation $Y = b_0 + b_1 * X$; in more detail: the variable Y can be expressed in terms of the constant (b_0) and the slope (b_1) multiplied by the variable X. The constant is sometimes also called the free term, and the slope is called the regression coefficient.

In the multidimensional case, when there is more than one independent variable, the regression line cannot be displayed in two-dimensional space; however, it can also be easily estimated. Then, in the general case, multiple regression procedures will evaluate the parameters of a linear equation of the form $Y = b_0 + b_1 * X_1 + b_2 * X_2 + \cdots + b_p * X_p$.

Regression coefficients represent the independent contributions of each independent variable to the prediction of the dependent variable. In other words, the variable X_1, for example, correlates with the variable Y after taking into account the influence of all other independent variables. This type of correlation is also referred to as *partial correlation*. The regression line expresses the best prediction of the dependent variable (Y) over the independent variables (X). However, nature rarely (if ever) is completely predictable and usually there is a significant scatter of the observed points relative to the fitted line (as was shown earlier in the scattering diagram). The deviation of an individual point from the regression line (from the predicted value) is called the *residual*.

The smaller the spread in the values of the residuals near the regression line with respect to the total spread in the values, the obviously better the forecast. For example, if there is no relationship between the variables X and Y, then the ratio of the residual variability of the variable Y to the original variance is 1.0. If X and Y are tightly coupled, then there is no residual variability, and the dispersion ratio will be 0.0. In most cases, the ratio will lie somewhere between these extreme values, i.e., between 0.0 and 1.0. This ratio is called the R-squared or *coefficient of determination*. This value is directly interpreted as follows. If there is an R-square equal to 0.4, then the variability of the values of the variable Y near the regression line is 1–0.4 of the initial variance; in other words, 40% of the original variability can be explained, and 60% of the residual variability remains unexplained. Ideally, it is desirable to have an explanation, if not for the whole, then at least for most of the original variability. The R-squared value is an indicator of the degree to which the model fits the data (an R-squared value close to 1.0 indicates that the model explains almost all the variability of the corresponding variables).

Usually, the degree of dependence of two or more predictors (independent variables or variables X) with the dependent variable (Y) is expressed using the *multiple correlation coefficient* R. By definition, it is equal to the square root of the coefficient of determination. This is a nonnegative value, taking values between 0 and 1. To interpret the direction of the relationship between the variables, look at the signs (plus or minus) of the *regression coefficients*. If the coefficient is positive, then the relationship of this variable with the dependent variable is positive; if the regression coefficient is negative, then the correlation is negative. Of course, if the regression coefficient is 0, there is no correlation between the variables.

Regression analysis is most common in comparison with other methods of multivariate analysis. It is used for a wide class of problems related mainly to the construction of a mathematical model of an object and its identification (Galvão, Araújo, & Soares, 2019; Pearce, 2009). When setting the task of selecting a model for describing patterns in an object, three characteristic situations are distinguished:

- The structure or type of model is known, and sample data are used to estimate coefficients or model parameters.
- Several models can be used to describe the object under study; it is necessary to choose the best of them and evaluate its parameters.
- The model of the studied object is unknown in advance; it is necessary to select, in the best way, some data from the sample data, the hypothesis of the structure of which can be put forward in the experiment.

The first two situations are quite definite and can be solved with various strategies for organizing sample data. Regarding the solution of the problem in conditions of model uncertainty, it is impossible to give a definite answer. The main difficulty is to put forward a realistic hypothesis regarding the proposed structure of the model. Experience shows that an unrealistic hypothesis leads to an inoperative model selected from sample data. The hypothesis regarding the structure of the model should be put forward on the basis of the greatest possible a priori information about the nature of the processes occurring in the object.

Only after the dimension of the problem is determined, independent variables that determine the studied patterns and dependencies in the object are selected, the suggested type of linearity of the model (by parameters, by variables) is clarified, and the possibility of introducing artificial effects in the object is evaluated, can we turn to the strategy of a passive experiment, which is based on regression analysis.

Let it be required to select a model of the object under study from the data collected in the mode of its normal functioning in the form of some function $\eta(X_1, X_2, \ldots, X_n)$ from the input variables X_i. The type of function h may not be known. Under these conditions, on the basis of existing knowledge about the object, it is advisable to find out the reality of the representation of this function as smooth and continuous. If such an opportunity exists for an object, then segments of a Taylor power series of various lengths obtained by expanding a smooth and continuous function in a neighborhood of the starting point in the space of input variables can be selected as a model. For example, for three variables, a function can have the form $\eta = \beta_1 X_1 + \beta_2 X_2 + \beta_3 X_3$ of a linear polynomial or, depending on the length of a segment of a series, polynomials of higher orders:

$$\eta = \beta_1 X_1 + \beta_2 X_2 + \beta_3 X_3 + \beta_{12} X_1 X_2 + \beta_{13} X_1 X_3 + \beta_{23} X_2 X_3 \\ + \beta_{123} X_1 X_2 X_3 + \cdots + \beta_{11} X_1^2 + \cdots. \tag{6.1}$$

Here, the coefficients of the polynomial are partial derivatives for the corresponding functions of the independent variables of the Taylor series, for example $\beta_3 = \frac{\partial_\eta}{\partial x_3}$, $\beta_{23} = \frac{\partial \eta^2}{\partial x_2 \partial x_3}$ and so on.

If the polynomial structure of the model is postulated, the situation in the selection of the model essentially reduces to the first two more defined situations in the statement of the problem. In other words, the task is reduced to estimating the parameters of the polynomial model and selecting the best regression set of input variables. The sample data for solving this problem contain N joint observations of the independent variables X_{ij} and the dependent variable Y_i, which represents the value of the function η_i, distorted by the influence of the error ε_i: $y_i = \eta_i(X_{ij}, \beta_j) + \varepsilon_i$, $i = 1, \ldots, N$, $j = 1, \ldots, k$.

Given the polynomial structure of the function η, we have

$$y_i = \beta_1 X_{i1} + \beta_2 X_{i2} + \cdots + \beta_k X_{ik} + \varepsilon_i. \tag{6.2}$$

The regression equation (6.1) is linear in the parameters b_j and nonlinear in the general case in the variables X_j. As was noted when considering ordinary least squares estimates, the independent variables in equation (6.2) can represent some function $\varphi(X)$ from observations, i.e., for example, $X_{ij} = X_{il} X_{iu}$, $X_{ij} = \ln X_{il}$, etc. In the case under consideration, in the form of segments of the Taylor series, the linearized variables in expression (6.2) replace the corresponding power functions of the original variables of expression (6.1). Further, for convenience, a linear notation of the regression equation is used. In matrix form, the regression equation is written as follows:

$$\mathbf{Y} = \mathbf{XB} + \mathbf{E}, \tag{6.3}$$

where $\mathbf{X} = \begin{bmatrix} X_{11} & X_{12} & \ldots & X_{1k} \\ X_{21} & X_{22} & \ldots & X_{2k} \\ \ldots & \ldots & \ldots & \ldots \\ X_{N1} & X_{N2} & \ldots & X_{Nk} \end{bmatrix}$; $\mathbf{Y} = \begin{bmatrix} y_1 \\ y_2 \\ \ldots \\ y_k \end{bmatrix}$; $\mathbf{B} = \begin{bmatrix} \beta_1 \\ \beta_2 \\ \ldots \\ \beta_k \end{bmatrix}$; $\mathbf{E} = \begin{bmatrix} \varepsilon_1 \\ \varepsilon_2 \\ \ldots \\ \varepsilon_N \end{bmatrix}$ are observation matrix of an independent variable, a column vector of observations of a dependent variable, column vectors of coefficients (parameters) of the regression equation, and random observation errors, respectively.

The use of regression analysis, which is based on the estimation of least squares, requires, in addition to the assumptions of least squares, the fulfillment of the assumption of a normal distribution of errors **E**.

If the vector **E** has an N-dimensional normal distribution with parameters 0, $\sigma^2 \mathbf{I}_N$, then the vector $\mathbf{Y} \in N_N(\mathbf{XB}, \sigma^2\mathbf{I}_N)$ (here $\mathbf{I}_N$ is the unit matrix). The estimates of the least-squares coefficients of equation (6.2) have the form:

$$\hat{\mathbf{B}} = \left(\mathbf{X}^T\mathbf{X}\right)^{-1}\mathbf{X}^T\mathbf{Y}, \tag{6.4}$$

where $\hat{\mathbf{B}} = \begin{bmatrix} b_1 \\ b_2 \\ \cdots \\ b_k \end{bmatrix}$ is column vector of the estimates of the coefficients β_j.

An unbiased estimate of the variance of errors σ^2 is the value of the residual variance relative to the regression equation, calculated by the formula:

$$\hat{\sigma}^2 = s_{res}^2 = \frac{\left(\mathbf{Y} - \mathbf{X}\hat{\mathbf{B}}\right)^T\left(\mathbf{Y} - \mathbf{X}\hat{\mathbf{B}}\right)}{N-k},$$

where k is the number of estimated coefficients of the equation.

An important role in the regression analysis, its modifications, and design criteria for the experiment, is played by the matrix $(\mathbf{X}^T\mathbf{X})^{-1}$. In the structure of the matrix of variance-covariances of OLS estimates $\mathbf{C} = s_{res}^2(\mathbf{X}^T\mathbf{X})^{-1}$ contains all the information about the accuracy and correlation properties of the estimates of the selected regression model.

The Fisher information matrix $\mathbf{M} = (1/N)(\mathbf{X}^T\mathbf{X})$ reflects the correlation structure of observations of input independent variables. Extracting the information contained in these matrices is the essence of the analysis of the regression model. For clarity, when analyzing the correlation properties of independent variables X_j, it is convenient to switch to normalized values of variables:

$$x_{ij} = \frac{X_{ij} - X_j}{s_{X_j}}, \quad j = 1, \ldots, k, \tag{6.5}$$

where X_j is the average value of the observations of the jth column of the matrix **X** and s_{X_j} is the standard deviation of the observations of the variable X_j:

$$s_{X_j}^2 = \frac{1}{N}\sum\nolimits_{i=1}^{N}\left(X_{ij} - \overline{X}_j\right)^2. \tag{6.6}$$

For normalized, variables, $M\{x_j\}=0$, $D\{x_j\}=1$. Then the $\mathbf{X}^T\mathbf{X}$ matrix will have the following form:

$$\mathbf{X}^{\mathrm{T}}\mathbf{X}=\begin{bmatrix} x_{11} & x_{12} & \ldots & x_{1k} \\ x_{21} & x_{22} & \ldots & x_{2k} \\ \ldots & \ldots & \ldots & \ldots \\ x_{N1} & x_{N2} & \ldots & x_{Nk} \end{bmatrix}\begin{bmatrix} x_{11} & x_{12} & \ldots & x_{1N} \\ x_{21} & x_{22} & \ldots & x_{2N} \\ \ldots & \ldots & \ldots & \ldots \\ x_{k1} & x_{k2} & \ldots & x_{kN} \end{bmatrix}$$

$$=\begin{bmatrix} \sum_{i=1}^{N} x_{i1}^2 & \sum_{i=1}^{N} x_{i1}x_{i2} & \ldots & \sum_{i=1}^{N} x_{i1}x_{ik} \\ \sum_{i=1}^{N} x_{i2}x_{i1} & \sum_{i=1}^{N} x_{i2}^2 & \ldots & \sum_{i=1}^{N} x_{i2}x_{ik} \\ \ldots & \ldots & \ldots & \ldots \\ \sum_{i=1}^{N} x_{ik}x_{i1} & \sum_{i=1}^{N} x_{ik}x_{i2} & \ldots & \sum_{i=1}^{N} x_{ik}^2 \end{bmatrix}=\begin{bmatrix} N & Nr_{12} & \ldots & Nr_{1k} \\ Nr_{12} & N & \ldots & Nr_{2k} \\ \ldots & \ldots & \ldots & \ldots \\ Nr_{1k} & Nr_{2k} & \ldots & N \end{bmatrix} \tag{6.7}$$

The diagonal elements of the matrix are equal to N, if we take into account (6.5), (6.6):

$$\sum_{i=1}^{N} x_{ij}^2=\frac{N}{N}\sum_{i=1}^{N}\frac{\left(X_{ij}-\overline{X}_j\right)^2}{s_{X_j}^2}=N.$$

The remaining elements of the matrix will be the correlation coefficients of the corresponding variables:

$$\sum_{i=1}^{N} x_{ij}x_{iu}=\frac{N}{N}\sum_{i=1}^{N}\frac{\left(X_{ij}-\overline{X}_j\right)\left(X_{iu}-\overline{X}_u\right)}{s_{X_j}s_{X_u}}=Nr_{ju}.$$

The Fisher information matrix will look like:

$$\mathbf{M}=\frac{1}{N}\mathbf{X}^T\mathbf{X}=\begin{bmatrix} 1 & r_{12} & \ldots & r_{1k} \\ r_{12} & 1 & \ldots & r_{2k} \\ \ldots & \ldots & \ldots & \ldots \\ r_{1k} & r_{2k} & \ldots & 1 \end{bmatrix}. \tag{6.8}$$

Representation of the matrix $\mathbf{X}^T\mathbf{X}$ in the form of Eq. (6.7) allows us to make an important observation. If the matrix $\mathbf{X}$ is provided with the property of orthogonality of the columns, i.e., $\sum_{i=1}^{N} x_{ij}x_{iu}=0$, then the $\mathbf{X}^T\mathbf{X}$ matrix will be diagonal with diagonal elements equal to N. The inverse matrix $(\mathbf{X}^T\mathbf{X})^{-1}$ will also be diagonal with $1/N$ diagonal elements. This property of orthogonality is one of the criteria for optimal planning in an active experiment.

Matrix $\mathbf{M}$ gives not only estimates of pair correlations, but also allows one to calculate estimates of multiple correlation coefficients using the formula:

$$R_{i0}^2 = \left(1 - \frac{|\mathbf{M}|}{|\mathbf{M}_{jj}|}\right), \tag{6.9}$$

where R_{j0}^2 is the square of the multiple correlation coefficient X_j with other variables, $|\mathbf{M}|$ is determinant of the matrix **M**, and $|\mathbf{M}_{jj}|$ is the determinant of the minor of the matrix **M**, obtained by deleting the jth row and jth column.

The dispersion-covariance matrix taking into account (6.8) can be written in the following form:

$$\hat{\mathbf{C}} = \begin{bmatrix} s^2\{b_1\} & c\hat{o}v\{b_1 b_2\} & \dots & c\hat{o}v\{b_1 b_k\} \\ c\hat{o}v\{b_2 b_1\} & s^2\{b_2\} & \dots & c\hat{o}v\{b_2 b_k\} \\ \dots & \dots & \dots & \dots \\ c\hat{o}v(b_k b_1) & c\hat{o}v\{b_k b_2\} & \dots & s^2\{b_k\} \end{bmatrix} = \frac{s_{res}^2}{N}\mathbf{M}^{-1}. \tag{6.10}$$

The variance of the estimates of the coefficients b_j can be obtained from (6.9), (6.10):

$$s^2\{b_j\} = \frac{s_{res}^2}{N}\frac{|\mathbf{M}_{jj}|}{|\mathbf{M}|} = \frac{s_{res}^2}{N\left(1 - R_{j0}^2\right)}. \tag{6.11}$$

In relation (6.11), one can see that with increasing tightness of the relationship between the independent variables at $R_{j0}^2 \to 1$, the accuracy of estimating the coefficients of the regression equation decreases, since $s^2\{b_j\} \to \infty$. This circumstance explains one of the reasons for the low performance of the regression models obtained in a passive experiment with a strong correlation of the input variables X_j.

Having received information about the accuracy of estimating the coefficients β_j, one can check the bullet hypotheses $H_0 : \beta_j = 0, j = 1, \dots, k$ about the significance of the coefficients of the regression model. For this, a selective distribution of t-statistics is used. Given that b_j are estimates of the least squares, that is, $M\{\beta_j\} = \beta_j$, and under the assumption that the hypothesis H_0: $\beta_j = 0$ is true, the t-statistic has the form:

$$t_j = \frac{|b_j|}{s_{res}}\sqrt{N\left(1 - R_{j0}^2\right)}.$$

If $t_j > t_{crit}[(N-k)(1 - {}^{\alpha}/_{2})]$, then the null hypothesis is incorrect, and the jth coefficient is considered significant.

Insignificant coefficients can be excluded from the regression equation with subsequent recalculation of all other coefficients. At this stage, sequential selection of the best set of independent regression variables is provided. In the simplest case, such a selection can be carried out according to the criterion of minimizing the residual scattering of the values of y_i relative to the selected model.

Great difficulties arise when trying to evaluate the performance of the model—its effectiveness in describing sample data. For this purpose, the relation $\gamma = \frac{s_{tot}^2\{y\}}{s_{res}^2}$ can be used where $s_{tot}^2 = \frac{1}{N-1}\sum_{i=1}^{N}(y_i - \overline{y})^2$, which shows how many times the total scattering of observations of the dependent variable Y occurs when using the selected model.

For different tasks, different g values can be considered effective. The experience of using regression models shows that the model is operational at $\gamma \geq 3$. A significant difference in the variance of σ_{tot}^2 and σ^2 can be revealed by testing hypothesis $H_1 : \sigma_{tot}^2 = \sigma^2$ relative to the competing hypothesis $H_0 : \sigma_{tot}^2 > \sigma^2$ with statistics $F_{calc} = s_{tot}^2 / s_{res}^2$.

If $F_{calc} > F_{crit}[(N-1)(N-k)(1-\alpha)]$, then hypothesis H_1 is valid, which indicates the achieved reduction in residual scattering of Y values when using the model to describe sample data.

In general, we can say that the linear regression method is widely used to solve the problems of numeric prediction in the energy sector, production, and, in particular, to ensure process safety and risk analysis in the process industry.

A wide range of tasks related to reliability arises in the repair and maintenance of aircraft. It is necessary to conduct a comprehensive analysis of failures by the frequency and duration of downtime, plan repairs, and calculate the number of spare parts in the warehouse (Le, Luo, Zhou, & Chan, 2015; Tianshan & Bo, 2016). In metallurgy and mechanical engineering, typical tasks associated with equipment failure are wearing of the rollers, destruction of the bearings, jamming of the bearings, etc.

In the energy sector, an urgent task is to predict the failure of elements, apparatus, and power transformers in distribution networks, depending on the factors affecting the months of the year. The obtained regression equations and functional dependences of the failure rate using the developed probabilistic forecasting models make it possible to estimate the number of spare elements, apparatus, and equipment for distribution networks and optimize these reserves taking into account the seasonality of operation, the principle of sufficiency, and determine the priority in financing funds for the formation of the reserve.

Another example is the effective operation of refinery equipment, which can be achieved by applying the latest performance and vibration analysis instruments (Balaam, 1977). The end result is reduced operating and maintenance costs by using preventive measures. It was pointed out that the key to dynamic predictive maintenance on reciprocating equipment is use of the analyzer to provide the following: accurate cylinder-by-cylinder fuel adjustment and horsepower balance; accurate determination of maintenance condition including valve losses; accurate load determination—pocket clearance on compressor cylinders and rotation speed can be optimized; and precise setting of ignition timing for maximum economy. The key to dynamic predictive maintenance of centrifugal equipment is the use of real time analysis to provide accurate determination of changes in maintenance condition—bearing condition, gear wear, looseness, etc.; precise measurement of any unbalance or misalignment; and operating data to avoid critical speeds or resonances, oil whip, oil whirl, and beat frequencies.

6.4 Advanced data analytics

Advanced methods of data analytics widely use intelligent algorithms and their combinations, and also make it possible to build predictive models and solve pattern recognition problems. Examples of such methods include time series analysis and text and image analysis.

6.4.1 Time series analysis

One area of application for machine learning and big data is the prediction of system behavior over time. To build forecasts, the concept of extrapolation and the associated time series analysis are used (Brockwell, 2010; Robinson, 2020).

Extrapolation is a special type of approximation, in which the function is approximated not between the given values, but outside the specified interval. In other words, extrapolation is an approximate determination of the values of the function $f(x)$ at points x lying outside the interval $[x_0, x_n]$, by its values at the points $x_0 < x_1 < \cdots < x_n$.

The most common extrapolation method is parabolic extrapolation, in which the value of $f(x)$ at x is taken to be the value of the polynomial $P_n(x)$ to the power n, taking the given values $y_i = f(x_i)$ at $n + 1$ point x_n. For parabolic extrapolation, interpolation formulas are used.

The general purpose of extrapolation is to extend the conclusions drawn from experiments on one part of a phenomenon to another part (in space or in time).

The initial data for extrapolation are time series. A *time series* (or a series of dynamics) is statistical material collected at different times about the value of any parameters (in the simplest case of one) of the process under study. Each unit of statistical material is called a *measurement* or a *reference*; it is also acceptable to call it a *level* at a point in time specified with it. In the time series, each report should indicate the measurement time or measurement number in order. The time series differs significantly from a simple data sample, since the analysis takes into account the relationship of measurements with time, and not just the statistical diversity and statistical characteristics of the sample (Fig. 6.12).

Time series analysis is a set of mathematical and statistical analysis methods designed to identify the structure of time series and to predict them. This includes, in particular, regression analysis methods. The identification of the structure of the time series is necessary in order to build a mathematical model of the phenomenon that is the source of the analyzed time series. The forecast of future values of the series is used for effective decision-making on risk management in the fields of environmental safety (Kumar, Kumar, & Kumar, 2020) and process safety (Nazaripour, Halvani, Jahangiri, Fallahzadeh, & Mohammadzadeh, 2018).

Time series consist of two elements:

- the period of time for which numerical values are given; and
- numerical values of one or another indicator, called the levels of the series.

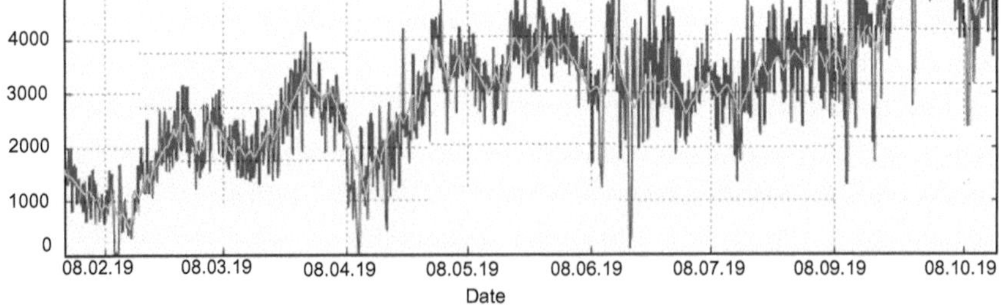

Fig. 6.12 Time series example.

Time series are classified according to the following criteria:

- According to the form of presentation of levels, there are rows of absolute indicators; relative indicators; average values.
- By the number of indicators for which levels are determined at each moment of time: one-dimensional and multidimensional time series.
- By the nature of the time parameter: moment and interval time series. In the instantaneous time series, the levels characterize the values of the indicator as of certain time points. In the interval series, the levels characterize the value of the indicator for certain periods of time. An important feature of the interval time series of absolute values is the possibility of summing their levels. Separate levels of the moment series of absolute values contain elements of repeated counting. This makes summation of the levels of moment series pointless.
- According to the type of interval between dates and time moments, time series can be classified as with equal intervals and unequal intervals.
- By the presence of missing values: complete and incomplete time series.
- Time series are deterministic and random: the first ones are obtained on the basis of the values of some nonrandom function (a series of sequential data on the number of days in months); the second is the result of the implementation of some random variable.
- Depending on the presence of the main trend, stationary series are distinguished in which the average value and variance are constant and nonstationary, containing the main development trend.

Time series, as a rule, arise as a result of measuring a certain indicator. These can be both indicators (characteristics) of technical systems, and indicators of natural, social, economic and other systems (for example, weather data). A typical example of a time series is the measurement of the parameters of a technological process, the analysis of which aims to determine the main direction (or trend) of development of a critical situation.

Depending on the type of random processes studied, for the analysis of time series, the so-called classical and special methods of analysis of time series are used (Brockwell, 2010; Robinson, 2020). By classical methods, we understand the methods that are developed for the analysis of *random stationary processes*, i.e., processes whose statistical properties do not change when the time reference is moved. Random stationary processes are found in radio engineering, communication theory, fluid and gas mechanics, oceanology, meteorology, etc.

The classical methods of analysis are conventionally divided into two groups in accordance with the goals achieved by processing the time series.

The first group includes probabilistic methods that are used to analyze and describe the statistical characteristics of a random process in the time domain. The main characteristics that are important for describing the statistical properties of random stationary processes are probability density, mean and variance, covariance, and correlation functions.

The second group includes parametric and nonparametric methods of spectral analysis, which are used to study the characteristics of a random process in the frequency domain. The main characteristic by which one can judge the spectral composition of the process under study is the spectral density function.

By special methods we mean methods that are developed for the analysis of nonstationary random processes, i.e., processes in which statistical properties change over time.

Most random processes encountered in practice are nonstationary in nature. Examples are unsteady waves in the ocean, unsteady geophysical processes, atmospheric and hydrodynamic turbulence, transients in radiophysical devices, information flows, etc. For the analysis of nonstationary random processes, classical methods turn out to be inapplicable. There is also no single methodology within which it is possible to study the properties of nonstationary random processes of various natures. Therefore, special methods have been developed designed to analyze only individual classes of processes. These include, in particular, methods based on the use of the Hilbert transform and the wavelet transform.

In some cases, nonstationary random processes corresponding to real physical phenomena have features that simplify their analysis and measurement: sometimes the data can be represented as a random process $X(t)$, all of whose sample functions have the form:

$$x(t) = a(t) + u(t) \tag{6.12}$$

or

$$x(t) = a(t)u(t), \tag{6.13}$$

where $u(t)$ is a sample function of a stationary random process and $a(t)$ is a deterministic function that defines the nonstationary part of the process, which is called a *trend*.

For sample functions of random processes discretized by argument, formulas (6.12), (6.13) take the form:

$$x(t_i) = a(t_i) + u(t_i), \tag{6.14}$$

$$x(t_i) = a(t_i)\,u(t_i), \tag{6.15}$$

where $u(t_i)$ is a stationary time series and $a(t_i)$ is a trend.

The possibility of representing sample functions in forms (6.14), (6.15) means that we can distinguish a trend in the time series as some deterministic component of the series that describes smooth or periodic changes in its characteristics (for example, average or variance). After the trend is removed, there remains only a random component of the time sequence, which has the property of stationarity. Obviously, if it is possible to carry out the operation of trend removal, then the analysis of nonstationary temporal implementation is greatly simplified and reduced to the study of its random stationary component by classical methods. That is why the analysis of time series of any nature must begin with the procedure of reducing to stationarity, and only if this is not possible is it advisable to apply special approaches.

Before starting the data analysis procedure, it is necessary to obtain this data, i.e., to perform a physical or numerical experiment, then classify the results of the experiment in order to determine methods for their further analysis.

The procedure for obtaining or collecting data is largely determined by the phenomenon being studied and the goals that are achieved by processing. Therefore, it is important to carry out preliminary planning of the experiment and determine its goals.

The first step in collecting data is to convert the process under investigation using special sensors into electrical form for further digitization of the data. At this stage, it is necessary to provide a linear conversion of the initial process into electrical form, as well as to calculate scale factors for the conversion of measurement results into physical units. In addition, it is necessary to minimize the level of extraneous noise caused by external conditions, as well as noise caused by recording equipment. The next step is to convert the data to digital form, i.e., obtaining time series, and, finally, an assessment of the basic properties and a detailed analysis of time series. Immediately before the experiment, it is necessary to conduct test experiments, including both data collection and processing (North, 2015).

Thus, the general data analysis scheme can be divided into three stages:

- preparation of data for analysis;
- assessment of the main properties of implementations; and
- data analysis itself.

Each of these steps requires a series of operations that are the subject of further consideration.

6.4.1.1 Preparation of data for analysis

The most important stage of the preliminary data analysis is the preparation of the collected material for detailed analysis. Research data can be presented in both discrete and continuous form. If the source data are continuous changes of any physical quantity, for example, the electric voltage taken from the sensors, then at this stage editing, sampling, and preprocessing are performed. If the data are the result of numerical experiments, i.e., presented in discrete form, the editing and discretization operations are not performed.

Editing is used prior to the analysis to identify and eliminate abnormal and distorted signals that may have occurred during data recording. The reason may be, for example, a high level of noise caused by environmental conditions or due to the use of amplifiers, a decrease in the signal level or its disappearance due to poor operation of the sensor, etc. Editing usually comes down to visual analysis of signal implementations before digitizing the data.

Discretization consists of converting an analog signal into digital form and consists of two operations that are not related to each other: actual discretization and quantization. Sampling itself is the process of determining the points in time at which samples should be taken; quantization is digitization of these samples. Typically, these two operations are carried out using analog-to-digital converters (ADCs), which connect the source of an analog signal to a computer.

At the preprocessing stage, a graphical representation and description of the behavior of the implementation of the random process, i.e., the construction and study of the chart of the time series, is provided. Graphical methods of analysis allow us to draw preliminary conclusions about the nature of the process, which can then be verified and refined by calculating the specific characteristics of the series. According to the time series chart, you can visually determine the presence of a trend and its nature. If the existence of a trend is obvious, i.e., unsteady series, it is necessary to isolate and remove the nonstationary component. The degree of smoothness or intermittent changes in the values of the series after the trend is removed makes it possible to judge the correlation between adjacent elements of the series. According to the chart, it is also possible to identify the presence of periodic components (Fig. 6.13).

A preliminary graphical analysis of the time series makes it possible to determine the presence of so-called *implausible values*, which are false and arise due to malfunctions in the data collection system. As a rule, implausible values are isolated points and differ greatly in magnitude from the average level of the time series or from its characteristic behavior.

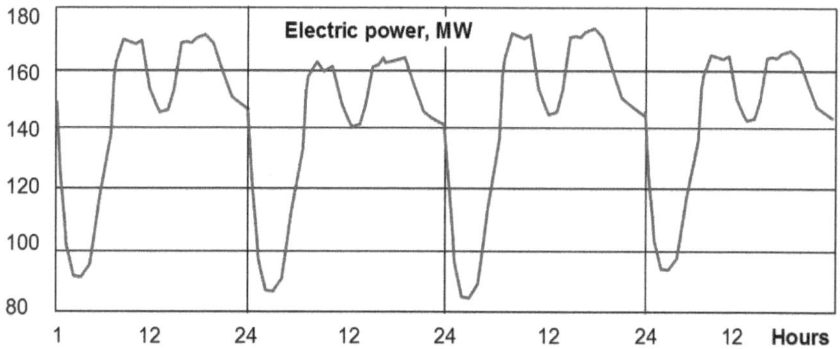

Fig. 6.13 An example of a time series with a periodic component.

In many cases, before evaluating the properties of the implementation, operations are carried out to bring to zero mean and unit variance.

The first operation is to evaluate the mean value $\mu(u)$ of the series $u(i)$ with the subsequent transition to a new time series $x(i)$ with a zero mean value:

$$x(i) = u(i) - \hat{\mu}(u). \tag{6.16}$$

This procedure is used to calculate almost all implementation characteristics. Therefore, it makes sense to implement it once, and not repeat it at each stage.

The second operation is carried out with a time series $x(i)$ and consists of estimating the standard deviation $\hat{\sigma}_x$ and normalizing the series:

$$y(i) = \frac{x(i)}{\hat{\sigma}_x}. \tag{6.17}$$

As a result, the normalized series $y(i)$ has unit dispersion. The transition to unit variance facilitates the construction of computational algorithms and simplifies the graphical representation of the characteristics of the time series.

6.4.1.2 Assessment of the main properties of implementations

The next important stage of the preliminary analysis is the assessment of the basic properties of the implementation of random processes. The main properties include *stationarity*, *normality*, and the presence of *periodic components*.

A preliminary assessment of the main characteristics makes it possible to simplify the study of the properties of time series; for example, identifying stationarity makes it possible to study the time series by classical methods, which are much simpler than methods for analyzing nonstationary realizations. If it is established that the time series contains periodic components, this avoids errors associated with incorrect interpretation of the analysis results.

6.4.1.3 Stationarity test

Methods of checking stationarity can be very different: from visual control of the time series by an experienced specialist to a detailed statistical assessment of the properties of random process implementations. In all cases, it is assumed that the time series correctly reflects the nature of the process being studied.

Obviously, of all the possible methods of analysis, the most valuable are formalized ones that can be used by nonspecialists in the field of analysis of random processes. These methods consist in checking the dependence (or independence) of the implementation parameters on the time reference. As parameters, as a rule, one selects the average value, variance, or less often, moment functions of higher orders and the probability density function. Depending on the selected characteristic, one speaks of the stationarity of the time series with respect to the average value or variance, etc. Note that the time series can be stationary with respect to one parameter, for example, the mean value, but exhibit nonstationarity with respect to another, for example, the variance.

Testing the time series $x(i)$ of length N for stationarity is carried out in the following order:

- The time series is divided into M equal intervals, and observations in different intervals are assumed to be independent.
- The estimates of the parameters of the series (mean value, variance, etc.) for each interval are calculated. These estimates form a sequence or time series of parameter estimates y_i; $1 \leq i \leq M$, for example, a series of mean values μ_i.
- The time series of estimates is checked for the presence of a trend or other changes in time, which cannot be explained only by selective variability of estimates. If the assessment trend exists, then the series is considered unsteady according to this assessment.

The basis for checking for the presence of a trend is the fact that for stationary implementation, estimates calculated at different intervals of the series are independent random variables. In other words, it is necessary to test for the statistical dependence between the elements of the time series of estimates of y_i.

Such a test can be carried out in various ways (including visual), which include both parametric and nonparametric criteria. Parametric criteria can be used if the frequency structure of the process is known. As a rule, such information is not available, therefore nonparametric criteria are used, for example, the inversion criterion, which is the most powerful tool for detecting monotone trends in time series and testing the hypothesis of statistical independence of observations.

Consider the procedure for constructing the inversion criterion for checking the statistical independence of the values $y_i, 1 \leq y \leq M$. The procedure consists of calculating how many times in the sequence the inequalities $y_k > y_i$ hold for $k < i$. Each such inequality is called an inversion.

Denote by A the total number of inversions. The formal procedure for calculating inversions is as follows. We define for the sequence y_i the quantities y_{ki}:

$$h_{ki} = \begin{cases} 1, & Y_k > Y_i, \quad k < i, \\ 0, & Y_k \leq Y_i. \end{cases} \tag{6.18}$$

Then the number of inversions is calculated as follows:

$$A = \sum_{k=1}^{M} A_k, \quad A_k = \sum_{i=k+1}^{M} h_{ki}. \tag{6.19}$$

More information about testing statistical hypotheses can be found in the literature on mathematical statistics (Finetti, 2017; Rasch, 2018).

In some cases, the independence of moment functions from the origin of the time can be insufficient evidence of stationarity, for example, if the spectrum of the time series is strongly unsteady. In this case, the time series in the frequency domain is divided into several adjacent frequency ranges using band-pass filters and the stationarity with respect to moment functions in each range is separately checked.

6.4.1.4 Checking the periodicity

Periodic and almost periodic components in the time series are diagnosed by the presence of delta peaks in the spectral density. However, in practice, difficulties often arise associated with finite frequency resolution of the power spectrum and effects of finite sample length. If the spectral density contains sharp peaks, then they can belong to both the periodic component of the time series and the narrow-band random component. If we calculate the power spectrum several times with a sequential increase in the frequency resolution, we can distinguish the harmonic signal from narrow-band noise.

For a harmonic signal, the width of the spectral peak decreases, and the height increases in proportion to the decrease in the width of the peak. For narrowband noise, the peak width may initially decrease, but, starting with some spectral resolution values, it does not change.

In order to implement this procedure for checking periodicity, it is necessary to change the spectral resolution in a wide range; this is done by changing the sampling frequency, but is not always possible. Therefore, a visual analysis of the probability density and correlation function, which are qualitatively different for the harmonic signal and noise, is additionally carried out. So, the probability density of a harmonic signal has two characteristic maxima, and the noise distribution is Gaussian; the correlation function of the noise process tends to zero with an increase in the time shift, and for a harmonic signal this function exhibits oscillations.

6.4.1.5 Normality test

The importance of this stage is due to the special place of the normal (Gaussian) probability density among various distribution functions: to find the parameters of a mathematical model of a time series with a Gaussian probability density, it is sufficient to estimate only the average value and variance of the series.

Checking the implementation of the joint venture for normality is carried out after it is found that the process is stationary, as well as that the periodic components are identified and excluded.

The easiest way to check for normality is to measure the probability density for the values of the time series and compare it with the normal theoretical distribution. If the length of the time series is large enough and the measurement errors are small compared with the deviations of the probability density from the normal curve, then the mismatch of the function with the normal distribution will be obvious. In addition, asymmetry and kurtosis can be calculated, the values of which for the Gaussian process are equal to $A=0$, $E=3\sigma^4$, respectively.

In doubtful cases, the normality test can be continued using the criteria of agreement, for example, the c χ-square test (Ergüner Özkoç, 2020; Robinson, 2020).

6.4.1.6 Data analysis

The analysis of the time series makes it possible to evaluate the main characteristics of the studied random process (Brockwell, 2010; North, 2015). The analysis scheme of the most important characteristics of random

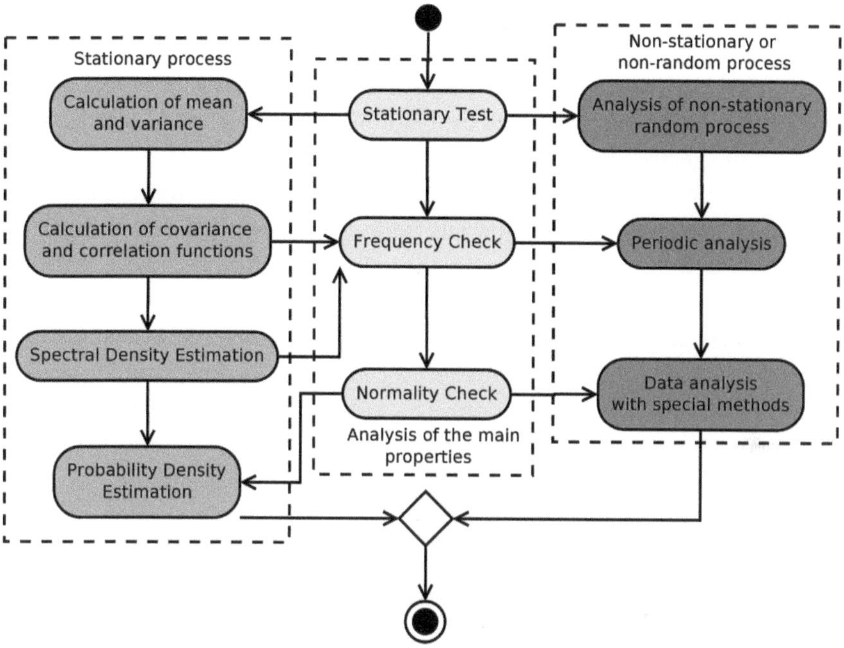

Fig. 6.14 Time series analysis scheme.

processes is presented in Fig. 6.14. Some of these steps may be excluded depending on the nature of the random processes being investigated and the research objectives.

The analysis of the basic properties is, in a sense, the stages of the preliminary analysis of experimental model data. This includes checking for stationarity, checking for periodicity, and checking for normality. The listed stages are necessary for revealing the basic statistical and spectral properties of SP implementations. They determine the choice of further methods for solving the analysis problem and minimizing the possible incorrect interpretation of the analysis results. These blocks were considered in the previous paragraph and included in the general scheme in order to understand better the relationship between the two stages of analysis—preliminary and actual data analysis.

Calculation of the mean and variance is required, firstly, due to the fact that these values determine the average level and spread of data, and, secondly, their estimation from short samples and using window functions allows us to study the process for stationarity.

The calculation of the covariance and correlation functions makes it possible to identify the presence or absence of periodic components in the time series at large time shifts, to evaluate the level of the noise component of the time series in the analysis of functions near zero. Note that if the implementation has a short duration, then the estimates of the covariance and correlation functions are characterized by a significantly lower error level than the estimate of spectral density.

The spectral density is estimated in order to find out the frequency composition of the time series and identify periodic components, which makes it possible to determine the energy relations between periodic and nonperiodic components, and also gives information about the nature of the time series and the properties of the object or process to which the series belongs.

The probability density estimation is carried out mainly in order to find out if the process is purely random, i.e., whether the probability density is described by one of the known distributions of random processes (Gaussian, Poisson, Rayleigh, exponential, etc.). If the probability density does not obey these laws, this indicates that the signal contains deterministic components, or the random process underwent a nonlinear transformation.

If as a result of the previous analysis it was established that the implementation under study belongs to a random nonstationary process, then it is necessary to continue the analysis of data by other methods. The choice of methods is determined by the objectives of the study. In other cases, it is preferable to reduce the time series to the stationary one, highlighting the nonstationary component. In addition, it is possible to apply directly to the nonstationary series analysis methods that are strictly applicable only to stationary data, but in this case one must be careful in interpreting the results.

If it is established that the implementation under study contains periodic or almost periodic components, then further analysis can be carried out in two directions. Firstly, it is possible to separate the random and periodic components and examine them separately. Secondly, it is possible to carry out their joint analysis, taking into account the presence of periodic components in the interpretation of the results. For example, if a spectrum containing harmonic components is constructed, then near the maxima of the spectral density at the frequencies corresponding to the harmonics, it is necessary to depict the delta functions. If this is not done, then one can give an incorrect interpretation of these spectral density maxima. The last block involves the use of special analysis methods, for example, the Hilbert transform or the wavelet transform, to obtain additional information about the properties of the time series or to solve particular problems determined by the objectives of the study.

There are also some special methods of time series analysis, for instance, clustering of time series data (Ergüner Özkoç, 2020).

Finally, we can say that time series analysis and forecasting have been widely recognized in various fields over years (Washington, Karlaftis, Mannering, & Anastasopoulos, 2020). It was recently applied for analyzing data on monitoring air quality. The most important air quality parameters are carbon monoxide (CO), carbon dioxide (CO_2), ammonia (NH_3), and acetone ($(CH_3)_2CO$). The sensor data was taken from three specific locations for predicting air quality the next day using linear regression as a machine learning algorithm. The model was evaluated using four performance indicators: mean absolute error (MAE), mean square error (MSE), mean square error (RMSE), and mean absolute percentage error (MAPE) (Kumar et al., 2020).

A short-term time series approach was also successfully implemented for safety performance evaluation in a steel industry (Nazaripour et al., 2018).

6.4.2 Text analysis

With the advent of big data, traditional methods of manually analyzing texts to identify key topics and trends in data are gradually losing their effectiveness. It is difficult to imagine that a team of process safety analysts daily receives and categorizes thousands of reports on incidents and hazards from hundreds of enterprises and branches, as well as the results of a process safety audit and risk assessment. Obviously, it is simply impossible to analyze manually all of these records in a reasonable timeframe. Software tools for text analytics can automate this process and increase its efficiency (Fairchild, 2016; Miner et al., 2012).

Text analytics, also often referred to as deep or intelligent text analysis, is an automated process for extracting important information from unstructured text data, in which methods are applied from various fields of knowledge, including computer linguistics, information retrieval, and statistics. Data analysts use text analytics tools to process employee survey results, records of call centers and employee negotiations, incident logs and their investigation results, etc. This can be useful in identifying the most common malfunctions and creating reports based on an analysis of call center call recordings, and identifying the main factors causing problems based on the analysis of incident reports.

Using natural language processing algorithms and statistical tools, text analytics allows you to solve problems such as classification of texts, analysis of tonality, and recognition of named entities and extraction of relationships (Aggarwal & Zhai, 2012; Do Prado, 2007). During the performance of these tasks, significant information is extracted from complex unstructured texts of large volume, which are thus converted into structured data. This allows companies to associate specific incidents with the effectiveness of the selected process safety management strategy.

Structuring such data allows analysts to quickly summarize and visualize trends in the data, which in turn leads to a better understanding of the data itself and the adoption of informed decisions on process safety management. As part of the text analysis, specialists have to perform several important tasks.

The primary concern is extracting the facts. During the analysis of text documents, the researcher often has to extract and structure the facts of interest. For example, during an explosion investigation at an oil refinery, analysts should study the journal entries and extract from them all the facts regarding the operation of the catalytic cracking unit: the names of the employees, the maintenance performed, the results of inspections, noted malfunctions, or deviations from normal operation.

The next task is the automatic coding of documents. When working with text documents, it is often necessary to identify special text patterns in them that correlate with predefined categories. This operation is called "document encoding." It is often used in the analysis of call center call recordings, survey data, comments and reviews, etc. The system combines text records in one category that express the same meaning, even though other lexical units are used. The coding methods are based on machine learning or a combination of technologies of linguistic and semantic analysis and recognition of text patterns.

Quite often, when the user is faced with the task of organizing a large number of documents, he or she has no idea either about the content or about the structure of these texts. In this situation, at the initial stage of the analysis, clustering of texts is usually performed, in which documents are distributed into groups depending on their general subject. The document clustering algorithm, based on hidden semantic analysis technology and a number of other advanced text analytics technologies, is widespread. To solve this problem, the *k*-means method can be adapted for clustering text documents into groups, where each group represents a collection of documents with a similar theme. The distance of the document to the centroid shows how closely the document speaks about this topic (Berry, 2003).

During the classification, documents are divided into groups depending on the main topic. The classification of texts is used, inter alia, to separate web pages and sites into thematic catalogs, fight against spam, and define the language of the text. Classification problems are important examples of using the naive Bayes classifier (Berry, 2003; Celardo & Everett, 2020).

Text annotation provides an opportunity to become familiar with brief annotations before reading the entire text, significantly increasing the speed and efficiency of analysts. Traditionally, the task of compiling brief annotations was assigned to employees of the company engaged in data analysis. Existing software systems offer tools and solutions that can read texts and automatically compose annotations for these texts by extracting some of the most important sentences from them.

Analyzing text in multiple languages is also an important task. In any country, companies most often have to analyze data in the language that the population of that country speaks. Also, this option allows you to process information on a regional and world scale. If you are faced with the task of translating all documents into one language, and then performing text analysis, you can use machine translation. However, it should be recognized that machine translation is not free from errors, especially when it comes to translating documents in Asian languages. Analysis of original documents, including elements of different languages, can give more accurate results.

Innovative solutions for the intellectual analysis of texts in natural languages with the aim of searching, extracting, and summarizing information about entities, facts, events, and their relationships include, for example, PROMT Analyzer SDK, Dandelion API, and PolyAnalyst.

The text analysis process usually consists of three important steps: parsing, searching and retrieval, and text mining.

Parsing is the process that uses unstructured text and prepares a structure for further analysis. The unstructured text could be a plain text file, a HyperText Markup Language (HTML) file, a Word document, or an Extensible Markup Language (XML) file. Parsing decomposes the provided text and displays it in a more structured way for the next steps.

Search and retrieval is the identification of the documents in a corpus that contain counting unique words (Fig. 6.15), and search items such as specific words, phrases, topics, or entities like people or organizations (Fig. 6.16). These search items are generally called key terms.

Text mining uses the terms and indexes produced by the previous two steps to discover meaningful insights pertaining to domains or problems of interest. Text mining is also known as intellectual text analysis. There is a growing interest in multilingual data mining: the ability to receive information in languages and to cluster similar elements from different linguistic sources in

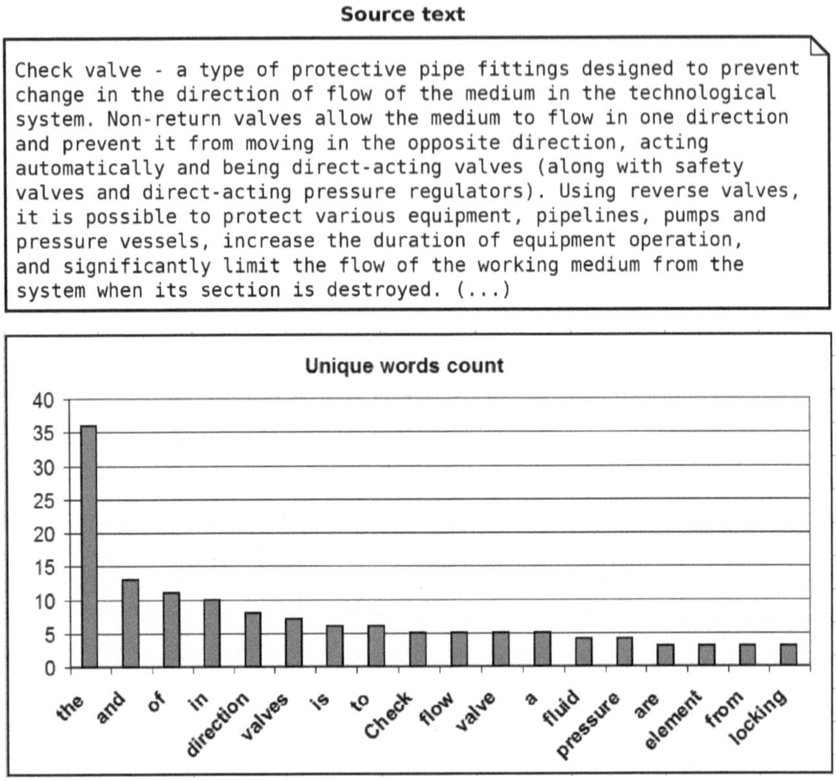

Fig. 6.15 Unique words count in a sample text.

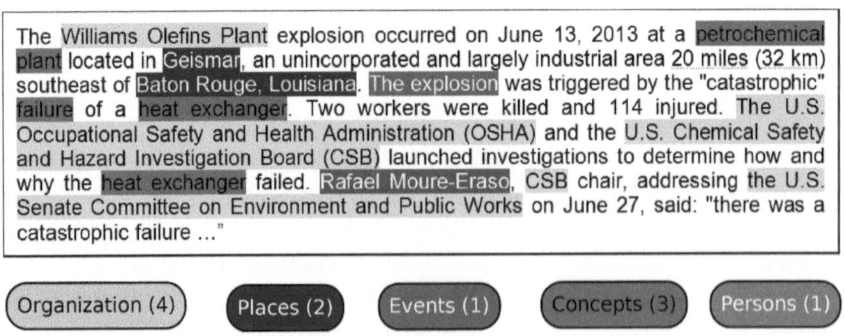

Fig. 6.16 Entity extraction in a sample text.

accordance with their value. Also relevant is the task of using a significant part of corporate information that comes in an "unstructured" form.

6.4.3 Image analysis

Image analysis is the extraction of meaningful information from images; mainly from digital images using digital image processing techniques. Image analysis tasks can be as simple as reading bar-coded tags or as complex as identifying a person by his or her face. Computers are indispensable for analyzing large amounts of data, for tasks requiring complex calculations, or for extracting quantitative information (Friel et al., 2000; Solomon & Breckon, 2010).

There are many different methods used for automatic image analysis. Each method can be useful for a small range of tasks; however, there are still no known methods of image analysis that would be universal enough for a wide range of tasks, compared with the ability to analyze human images. Examples of image analysis methods in various fields:

- image segmentation;
- 2D and 3D object recognition;
- motion detection;
- video tracking;
- number plate recognition;
- pedestrian and transportation flows analysis (Tripodi et al., 2020); and
- medical image processing.

In the field of industrial safety, as examples of the use of image analysis, one can consider checking the presence of a hardhat on the head of an employee and recognizing the fact of smoking in images and videos. When solving the problems of compliance analysis with personnel safety requirements during hazardous work, image analysis systems based on video sensors and convolutional neural networks are used.

Using these technologies, for example, it is possible to:

- check personnel and equipment during technological operations, for example, the presence of a protective hardhat on the head of an employee; and
- detect violations of safety regulations by an employee on the premises of the enterprise, for example, smoking in prohibited places in the vicinity of hazardous products according to safety regulations in the oil industry.

Let's look at these two examples in terms of the features of using intelligent technologies and big data for pattern recognition. Both tasks are reduced to the task of classifying recognized patterns. In other words, you first need to identify people in the image, and then classify these images and highlight objects with abnormal behavior. In the first case, not wearing a hardhat will be considered abnormal behavior, and in the second case, the fact of smoking.

Hardhats are an important safety tool used to protect industry and construction workers from accidents. However, injuries caused by not wearing hardhats still happen. In order to increase the efficiency of supervision of industrial and construction workers to prevent injuries, automatic nonhardhat-use detection technology can be very important. Existing automatic methods of detecting hardhats are commonly limited to the detection of objects in near-field surveillance videos. The proposed methods use of a high precision, high speed, and widely applicable Faster R-CNN (Region-based Convolutional Network) and some other methods to detect construction workers' nonhardhat-use (Fig. 6.17) (Kim, Nam, Kim, & Cho, 2009; Luo & Wang, 2019; Park, Elsafty, & Zhu, 2015).

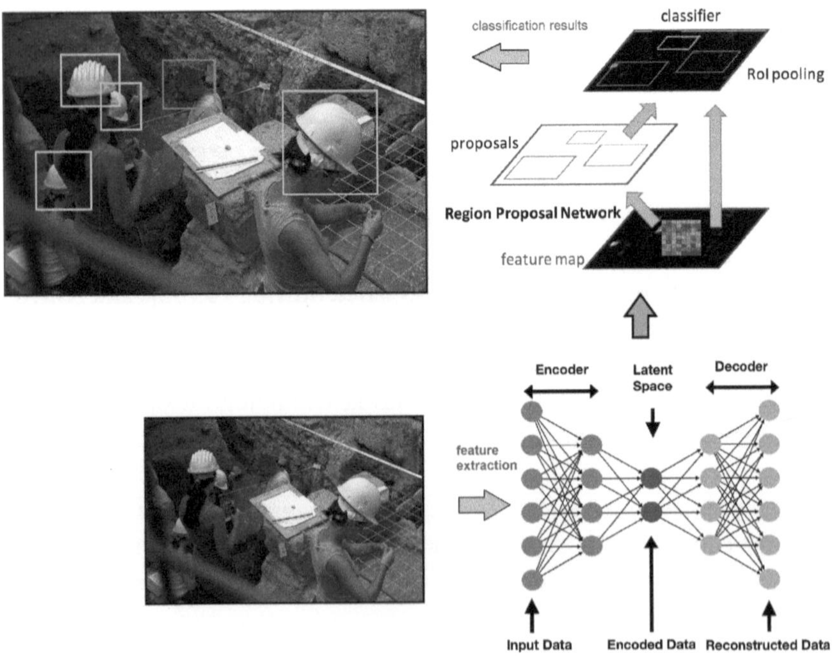

Fig. 6.17 Hardhat-wearing detection based on deep learning algorithms.

Convolutional neural networks (CNN) have been used for various computer vision tasks such as image classification, object detection, semantic segmentation, face recognition, and hyperspectral image recognition, among others. In particular, in the field of object detection, networks such as Faster R-CNN, R-FCN, YOLO, and SSD have achieved high accuracy and, therefore, have attracted the attention of researchers. Faster R-CNNs and R-FCNs are two-stage networks consisting of application generation processes in a region and application classification.

To evaluate the performance of Faster R-CNN, more than 100,000 worker image frames were randomly selected from the far-field surveillance videos of different sites over a period of more than a year. The research analyzed various visual conditions of the sites and classified image frames according to their visual conditions. The image frames were input into Faster R-CNN according to different visual categories. The experimental results demonstrate that the high precision, high recall, and fast speed of the method can effectively detect workers' nonhardhat-use in different conditions, and can facilitate improved safety inspection and supervision.

Another method is hardhat-wearing detection based on a lightweight convolutional neural network with multiscale features and a top-down model (Wang et al., 2020).

The proposed convolutional neural network can solve the problem of determining whether work hardhats are worn. The proposed method is aimed at identifying the human head and determining whether it is wearing a hardhat. The MobileNet model is used as a backbone network, which allows the detector to work in real time. A downstream module is used to improve the process of retrieving objects. Finally, heads with hardhats and without hardhats were detected on multiscale objects using a prediction module based on residual blocks. The experimental results for the data set that we established show that the proposed method can give an average accuracy of 87.4%/89.4% at a speed of 62 frames per second for detecting people without/with hardhats on.

The detection of abnormal behavior in crowded scenes is extremely difficult in the field of computer vision due to serious occlusions between objects, different crowd density, and the complex mechanics of the human crowd. The most effective are the methods for detecting and localizing abnormal actions in video sequences of scenes with human learning. A key feature of such methods is the combination of anomaly detection with space-time convolutional neural networks (CNN) (Chiu & Kuo, 2018; Sadek, Al-Hamadi, Michaelis, & Sayed, 2010). Such an architecture makes it possible to capture elements from both spatial and temporal dimensions by performing spatial-temporal convolutions, thereby extracting both appearance information and motion information encoded in continuous frames. Spatial-temporal convolutions are performed only in the spatial-temporal volumes of moving pixels in order to provide resistance to local noise and improve detection accuracy. An experimental evaluation of the model was carried out on reference data sets containing various situations with crowds of people, and the results demonstrate that the proposed approach is superior to other methods. Spatial-temporal convolutional neural network is designed to highlight automatically the spatial-temporal characteristics of the crowd. Anomaly detection performance improves when analysis focuses only on dynamic areas. Anomaly events that occur in small regions are effectively detected and localized by a spatial-temporal convolutional neural network (Fig. 6.18).

The proposed algorithm consists of frame extracting, preprocessing stage, training stage, and testing stage. In Fig. 6.18, the TSN segment is a segment of the temporal segment network (TSN), a novel framework for video-based action recognition. It refers to the idea of long-range temporal structure modeling. It includes a sparse temporal sampling and video-level supervision. Data augmentation is a common technique to improve results and avoid overfitting during model training.

One study suggested a system that can recognize smoking. It uses data balancing and augmentation based on GoogLeNet network architecture and temporary networks to achieve effective smoking recognition. Experimental results show that the smoking accuracy rate can reach 100% for the Hmdb51 test data set (Chiu & Kuo, 2018).

6.5 Summary

Data analytics is an important part of the risk management system in which data is organized, processed, and analyzed, and the extraction and preparation of analytical data are presented in the form of graphs, charts, and diagrams.

Machine learning is a class of artificial intelligence methods that is not a direct solution to problems. To implement such methods, mathematical statistics, numerical methods, measurement methods, probability theory, graph theory, and various methods of working with data in digital form are used. Machine learning algorithms can be divided into categories according to their purpose. Understanding the categories of learning algorithms and the ability to choose the right category is an important step in using data to manage process safety risks.

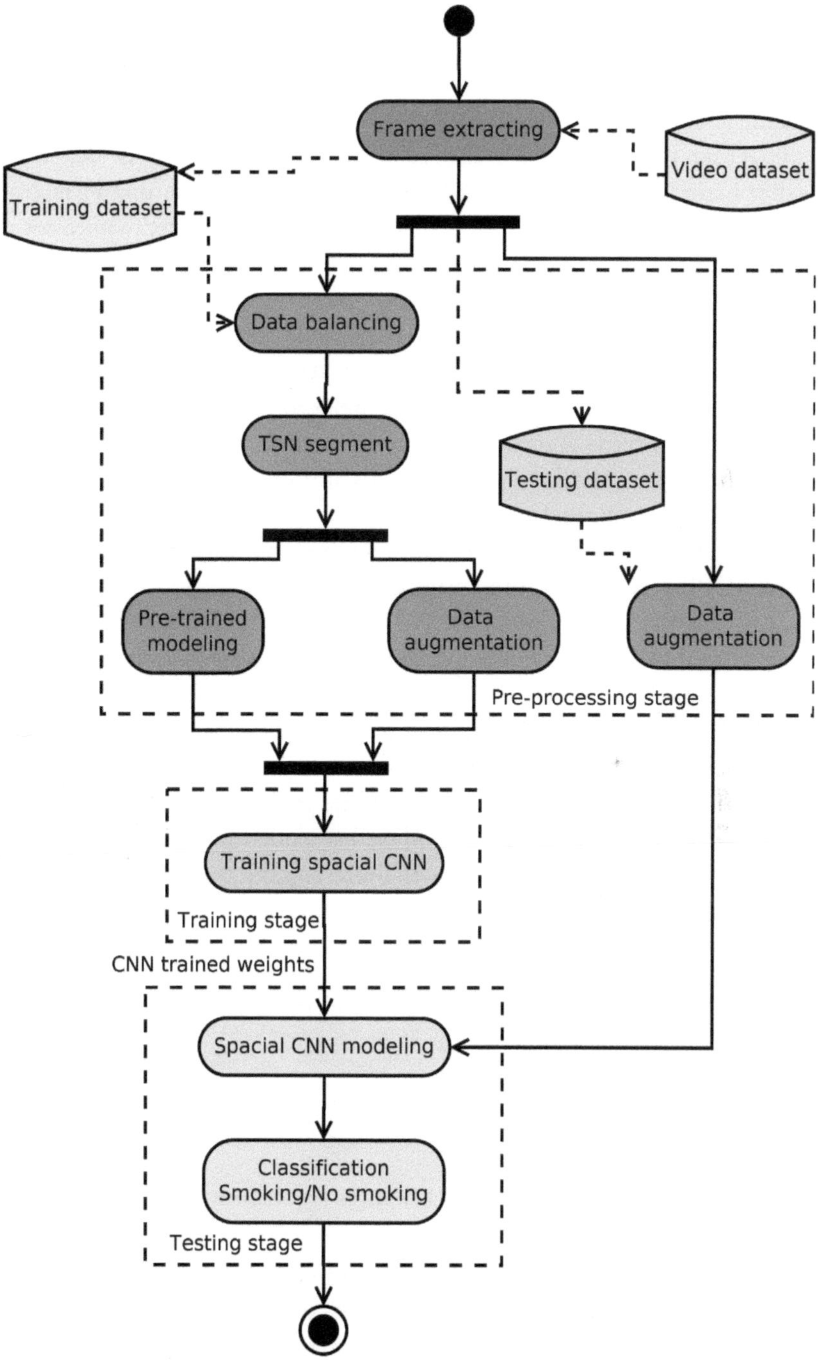

Fig. 6.18 Smoking action recognition based on spatial-temporal convolutional neural networks.

Clustering is a method often used for exploratory analysis of the data. In clustering, there are no predictions made. Rather, clustering methods find the similarities between objects according to the object features and group the similar objects into clusters.

Together with analytical techniques such as clustering, classification is another basic learning method that is used to solve data mining problems. When classification learning, the classifier is presented along with a set of examples that are already classified. The main task solved by the classifiers is to assign class labels to new observations or examples.

Predictive modeling is primarily concerned with minimizing model errors or, in other words, forecasting as accurately as possible. The machine learning borrows algorithms from various fields, including statistics, and uses them for this purpose. Linear regression is perhaps one of the most well-known and understandable algorithms in statistics and machine learning.

One area of application for machine learning and big data is the prediction of system behavior over time. To build forecasts, the concept of extrapolation and the associated time series analysis are used.

Text analytics, also often referred to as deep or intelligent text analysis, is an automated process for extracting important information from unstructured text data, in which methods are applied from various fields of knowledge, including computer linguistics, information retrieval, and statistics. Image analysis is the extraction of meaningful information from images, mainly from digital images using digital image processing techniques. Image analysis tasks can be as simple as reading bar-coded tags or as complex as identifying a person by his or her face. When solving the problems of compliance analysis with personnel safety requirements during hazardous work, image analysis systems based on video sensors and convolutional neural networks are used.

6.6 Definitions

Data analytics is the science of analyzing raw data in order to make conclusions about that information.

Machine learning is the study of computer algorithms that improve automatically through experience. It is seen as a subset of artificial intelligence.

Predictive modeling is the process of using known results to create, process, and validate a model that can be used to forecast future outcomes.

Descriptive models relate to the recording and analysis of statistical data to enhance the capabilities of business intelligence.

Pattern recognition is the automated recognition of patterns and regularities in data.

Cluster analysis is the task of grouping many objects into subsets (clusters) so that objects from one cluster are more similar to each other than objects from other clusters by some criterion.

Metalearning is a subfield of machine learning where automatic learning algorithms are applied on metadata about machine learning experiments.

Classification is the problem of identifying to which of a set of categories (subpopulations) a new observation belongs, on the basis of a training set of data containing observations (or instances) whose category membership is known.

Linear regression is a linear approach to modeling the relationship between a scalar response (or dependent variable) and one or more explanatory variables (or independent variables).

A time series is a series of data points indexed (or listed or graphed) in time order.

Text mining is the task of automatically extracting structured information from unstructured and/or semistructured machine-readable documents and other electronically represented sources.

Image analysis is the extraction of meaningful information from images; mainly from digital images using digital image processing techniques.

References

Aggarwal, C. C., & Zhai, C. (2012). *Mining text data*. Springer.

Albalawi, F., Durand, H., & Christofides, P. D. (2017). Distributed economic model predictive control for operational safety of nonlinear processes. *AICHE Journal*, *63*(8), 3404–3418. https://doi.org/10.1002/aic.15710.

Al-Hadraawy, M. (2020). *Machine learning*. https://doi.org/10.13140/RG.2.2.15016.72969.

Angelini, C. (2018). Regression analysis. In *Vol. 1–3. Encyclopedia of bioinformatics and computational biology: ABC of bioinformatics* (pp. 722–730). Italy: Elsevier. https://doi.org/10.1016/B978-0-12-809633-8.20360-9.

Balaam, E. (1977). Dynamic predictive maintenance for refinery equipment. *Hydrocarbon Processing*, *56*(5), 131–136.

Berrar, D. (2018). Bayes' theorem and naive bayes classifier. In *Vol. 1–3. Encyclopedia of bioinformatics and computational biology: ABC of bioinformatics* (pp. 403–412). Japan: Elsevier. https://doi.org/10.1016/B978-0-12-809633-8.20473-1.

Berry, M. W. (Ed.). (2003). *Survey of text mining I: Clustering, classification, and retrieval* Springer.

Bikmukhametov, T., & Jäschke, J. (2020). Combining machine learning and process engineering physics towards enhanced accuracy and explainability of data-driven models. *Computers & Chemical Engineering*. https://doi.org/10.1016/j.compchemeng.2020.106834, 106834.

Brockwell, P. J. (2010). Time series analysis. In *International encyclopedia of education* (pp. 474–481). United States: Elsevier Ltd. https://doi.org/10.1016/B978-0-08-044894-7.01372-5.

Carbonell, J. G., Michalski, R. S., & Mitchell, T. M. (1983). An overview of machine learning. In *Machine learning*. https://doi.org/10.1016/B978-0-08-051054-5.50005-4 (chapter 1).

Caruana, R., & Niculescu-Mizil, A. (2006). An empirical comparison of supervised learning algorithms. In *ICML 2006—Proceedings of the 23rd international conference on machine learning, United States*.

Carvalko, J. R., & Preston, K. (1972). *On determining optimum simple Golay marking transforms for binary image*.

Celardo, L., & Everett, M. G. (2020). Network text analysis: A two-way classification approach. *International Journal of Information Management, 51*. https://doi.org/10.1016/j.ijinfomgt.2019.09.005.

Celebi, M. E., Kingravi, H. A., & Vela, P. A. (2013). A comparative study of efficient initialization methods for the k-means clustering algorithm. *Expert Systems with Applications, 40*(1), 200–210. https://doi.org/10.1016/j.eswa.2012.07.021.

Chan, L., & Tang, W. (2019). *Big data analytics* (pp. 113–121). Informa UK Limited.

Chiu, C.-F., & Kuo, C.-H. (2018). *Smoking action recognition based on spatial-temporal convolutional neural networks* (pp. 1616–1619).

Do Prado, H. A. (2007). *Emerging technologies of text mining: Techniques and applications*. Idea Group Reference.

Domingos, P., & Pazzani, M. (1997). On the optimality of the simple Bayesian classifier under zero-one loss. *Machine Learning, 29*(2), 103–137.

Ergüner Özkoç, E. (2020). *Clustering of time-series data*. IntechOpen. https://doi.org/10.5772/intechopen.84490.

Fairchild, T. E. (2016). Technology: Computerized text analysis. In *The curated reference collection in neuroscience and biobehavioral psychology*. Japan: Elsevier Science Ltd. https://doi.org/10.1016/B978-0-12-809324-5.23731-0.

Finetti, B. (2017). *Mathematical statistics*. https://doi.org/10.1002/9781119286387.ch12.

Fratello, M., & Tagliaferri, R. (2018). Decision trees and random forests. In *Vol. 1–3. Encyclopedia of bioinformatics and computational biology: ABC of bioinformatics* (pp. 374–383). Italy: Elsevier. https://doi.org/10.1016/B978-0-12-809633-8.20337-3.

Friel, J. J., et al. (2000). *Practical guide to image analysis*. ASM International.

Galvão, R. K. H., de Araújo, M. C. U., & Soares, S. F. C. (2019). Linear regression modeling: Variable selection, reference module in chemistry. Molecular sciences and chemical engineering. Elsevier B.V.. doi:https://doi.org/10.1016/b978-0-12-409547-2.14706-7.

Ganesh, S. (2010). Multivariate linear regression. In *International encyclopedia of education* (pp. 324–331). New Zealand: Elsevier Ltd. https://doi.org/10.1016/B978-0-08-044894-7.01350-6.

Goel, P., Datta, A., & Sam Mannan, M. (2017). Application of big data analytics in process safety and risk management. In *Proceedings—2017 IEEE international conference on big data, big data 2017*. United States: Institute of Electrical and Electronics Engineers Inc. https://doi.org/10.1109/BigData.2017.8258040.

Goel, P., Pasman, H., Datta, A., & Mannan, S. (2019). How big data & analytics can improve process and plant safety and become an indispensable tool for risk management. *Chemical Engineering Transactions*, 77, 757–762. https://doi.org/10.3303/CET1977127.

Goel, P., Datta, A., & Mannan, S. (2017). Industrial alarm systems: Challenges and opportunities. *Journal of Loss Prevention in the Process Industries*, 23–36. https://doi.org/10.1016/j.jlp.2017.09.001.

Golden, R. M. (2015). Statistical pattern recognition. In *International encyclopedia of the social & behavioral sciences* (2nd ed., pp. 411–417). United States: Elsevier Inc. https://doi.org/10.1016/B978-0-08-097086-8.43093-X.

Gori, M. (2018). *Machine learning: A constraint-based approach*. https://doi.org/10.1016/C2015-0-00237-4.

Gudivada, V. N., Irfan, M. T., Fathi, E., & Rao, D. L. (2016). Cognitive analytics: Going beyond big data analytics and machine learning. In *Vol. 35. Handbook of statistics* (pp. 169–205). United States: Elsevier B.V. https://doi.org/10.1016/bs.host.2016.07.010.

Haesik, K. (2020). *Machine learning* (pp. 151–193). Wiley.

John, G. H., & Langley, P. (1995). Estimating continuous distributions in Bayesian classifiers. In *Proc. eleventh conf. on uncertainty in artificial intelligence*.

Kim, M., Nam, Y., Kim, S., & Cho, W. (2009). A hardhat detection system for preventing work zone accidents in complex scene images. In *Presented at the conference: Proceedings of the 2009 international conference on image processing, computer vision, & pattern recognition, IPCV 2009, Las Vegas, NV, USA*.

Kotu, V., & Deshpande, B. (2019). *Data science: Concepts and practice* (2nd). Morgan Kaufmann (an imprint of Elsevier).

Kriegel, H.-P., Schubert, E., & Zimek, A. (2017). The (black) art of runtime evaluation: Are we comparing algorithms or implementations? *Knowledge and Information Systems*, 341–378. https://doi.org/10.1007/s10115-016-1004-2.

Kumar, R., Kumar, P., & Kumar, Y. (2020). Time series data prediction using IoT and machine learning technique. *Procedia Computer Science*, 373–381. https://doi.org/10.1016/j.procs.2020.03.240.

Lantz, B. (2019). *Machine learning with R* (3rd ed.). Packt Press.

Le, T., Luo, M., Zhou, J., & Chan, H. (2015). Predictive maintenance decision using statistical linear regression and kernel methods. In *19th IEEE international conference on emerging technologies and factory automation, ETFA 2014*. https://doi.org/10.1109/ETFA.2014.7005357.

Lemke, C., Budka, M., & Gabrys, B. (2013). Metalearning: A survey of trends and technologies. *Artificial Intelligence Review*, *44*(1). https://doi.org/10.1007/s10462-013-9406-y. ISSN 0269-2821.

Luo, W., & Wang, Q. (2019). Hardhat-wearing detection with cloud-edge collaboration in power internet-of-things. In *Proceedings—2019 4th international conference on mechanical, control and computer engineering, ICMCCE 2019*. China: Institute of Electrical and Electronics Engineers Inc. https://doi.org/10.1109/ICMCCE48743.2019.00158.

MacQueen, J. (1967). Some methods for classification and analysis of multivariate observations. In *Proceedings of the fifth Berkeley symposium on mathematical statistics and probability*. Berkeley, CA: University of California Press.

Maron, M. E. (1961). Automatic indexing: An experimental inquiry. *Journal of the ACM*, *8*(3), 404–417. https://doi.org/10.1145/321075.321084.

Miner, G., Elder, J., Hill, T., Nisbet, R., Delen, D., & Fast, A. (2012). *Practical text mining and statistical analysis for non-structured text data applications*. Elsevier BV.

Nagel, W., & Ludwig, T. (2019). *Data analytics. Informatik. Vol. 42*. Spektrum. https://doi.org/10.1007/s00287-019-01231-9.

Nazaripour, E., Halvani, G., Jahangiri, M., Fallahzadeh, H., & Mohammadzadeh, M. (2018). Safety performance evaluation in a steel industry: A short-term time series approach. *Safety Science*, 285–290. https://doi.org/10.1016/j.ssci.2018.08.028.

North, G. R. (2015). Statistical methods: Data analysis: Time series analysis. In *Encyclopedia of atmospheric sciences* (2nd ed., pp. 205–210). United States: Elsevier Inc. https://doi.org/10.1016/B978-0-12-382225-3.00131-6.

Park, M. W., Elsafty, N., & Zhu, Z. (2015). Hardhat-wearing detection for enhancing on-site safety of construction workers. *Journal of Construction Engineering and Management*, *141*(9). https://doi.org/10.1061/(ASCE)CO.1943-7862.0000974.

Pearce, J. (2009). *Regression: Linear and nonlinear* (pp. 373–379). Elsevier B.V.

Pelleg, D., & Moore, A. (1999). Accelerating exact k-means algorithms with geometric reasoning. In D. Pelleg, & A. Moore (Eds.), *Presented at the proceedings of the fifth ACM SIGKDD international conference on knowledge discovery and data mining—KDD'99*. San Diego, CA: ACM Press.

Pimentel, B. A., & de Carvalho, A. (2020). A meta-learning approach for recommending the number of clusters for clustering algorithms. *Knowledge-Based Systems, 195*.

Rasch, D. (2018). *Mathematical statistics*. Wiley.

Robinson, G. M. (2020). Time series analysis. In *International encyclopedia of human geography*. https://doi.org/10.1016/B978-0-08-102295-5.10614-6.

Sadek, S., Al-Hamadi, A., Michaelis, B., & Sayed, U. (2010). Toward robust action retrieval in video. In *British machine vision conference, BMVC 2010—Proceedings*. Germany: British Machine Vision Association, BMVA. https://doi.org/10.5244/C.24.44.

Samson, D. (2019). *Analytics*. https://doi.org/10.1201/9781315270883-7.

Schaul, T., & Schmidhuber, J. (2010). Metalearning. *Scholarpedia*, 4650. https://doi.org/10.4249/scholarpedia.4650.

Sedgwick, P. (2014). Unit of observation versus unit of analysis. *BMJ, 348*. https://doi.org/10.1136/bmj.g3840.

Solomon, C. J., & Breckon, T. P. (2010). *Fundamentals of digital image processing*. Wiley-Blackwell.

Steven, D., Brown, A. J., & Myles. (2019). Decision tree modeling, reference module in chemistry. In *Molecular sciences and chemical engineering*. https://doi.org/10.1016/B978-0-12-409547-2.00653-3. ISBN: 9780124095472.

Theodoridis, S. (2015). *Machine learning*. https://doi.org/10.1016/B978-0-12-801522-3.00001-X.

Tianshan, G., & Bo, G. (2016). *Failure rate prediction of substation equipment combined with grey linear regression combination model* (pp. 1–5).

Trevino, A. (2016). *Introduction to K-means clustering*. Retrieved 3 May 2020, from https://blogs.oracle.com/datascience/introduction-to-k-means-clustering.

Tripodi, A., Mazzia, E., Reina, F., Borroni, S., Fagnano, M., & Tiberi, P. (2020). A simplified methodology for road safety risk assessment based on automated video image analysis. *Transportation Research Procedia*, 275–284. https://doi.org/10.1016/j.trpro.2020.03.017.

Uttal, W. R. (2002). Pattern recognition. In *Encyclopedia of the human brain* (pp. 787–795). Elsevier B.V. https://doi.org/10.1016/b0-12-227210-2/00270-3.

Wang, L., Xie, L., Yang, P., Deng, Q., Du, S., & Xu, L. (2020). Hardhat-wearing detection based on a lightweight convolutional neural network with multi-scale features and a top-down module. *Sensors*, 1868. https://doi.org/10.3390/s20071868.

Washington, S., Karlaftis, M., Mannering, F., & Anastasopoulos, P. (2020). *Forecasting in time series* (pp. 173–195). Informa UK Limited.

Witten, I. H., Frank, E., Hall, M. A., & Pal, C. J. (2017). Algorithms: The basic methods. In *Data mining* (4th ed., pp. 91–160). Morgan Kaufmann (chapter 4).

Zhang, Y.-C., & Sakhanenko, L. (2019). The naive Bayes classifier for functional data. *Statistics & Probability Letters*, 137–146. https://doi.org/10.1016/j.spl.2019.04.017.

CHAPTER 7

Risk control and process safety management systems

7.1 Hierarchical safety management system

To solve the problems of control and risk management, to ensure the safety of technological processes, it is necessary to use all available information generated in real time. This allows us to use the developed risk minimization scenarios in the event of critical situations to predict the development of the course of the critical situation.

The solution to this large set of problems is possible using a hierarchical approach to organizing risk management processes, data collection, and processing technologies based on modern computer technologies.

The procedures for the formation of the necessary data flows and their use at various levels of the risk management system are presented in Fig. 7.1. It should be noted that the solution of tasks to minimize risks requires the implementation of many simultaneously performed procedures, which requires special control by staff and process safety managers.

Let us consider in more detail the principles of organizing a hierarchical system of minimizing risks and methods of processing information at various levels of risk management.

Many technological operations of the production process are interconnected; changing the parameters of one operation affects the progress of another. Managing many parameters of a large system is very difficult, and in the event of a dangerous situation or failure of a subsystem, the control task is complicated many times over. For the development of control systems for large systems, the principles of designing large systems are applied, which allow solving the problem based on the decomposition of the system into subsystems. In this case, a control system is developed for each subsystem taking into account their interaction.

The main design principles in solving the problem of managing a large system include the following:

- The principle of functional integration. This principle implies that the creation of an integrated control system is based on the coordination of

Process Safety and Big Data
https://doi.org/10.1016/B978-0-12-822066-5.00005-4

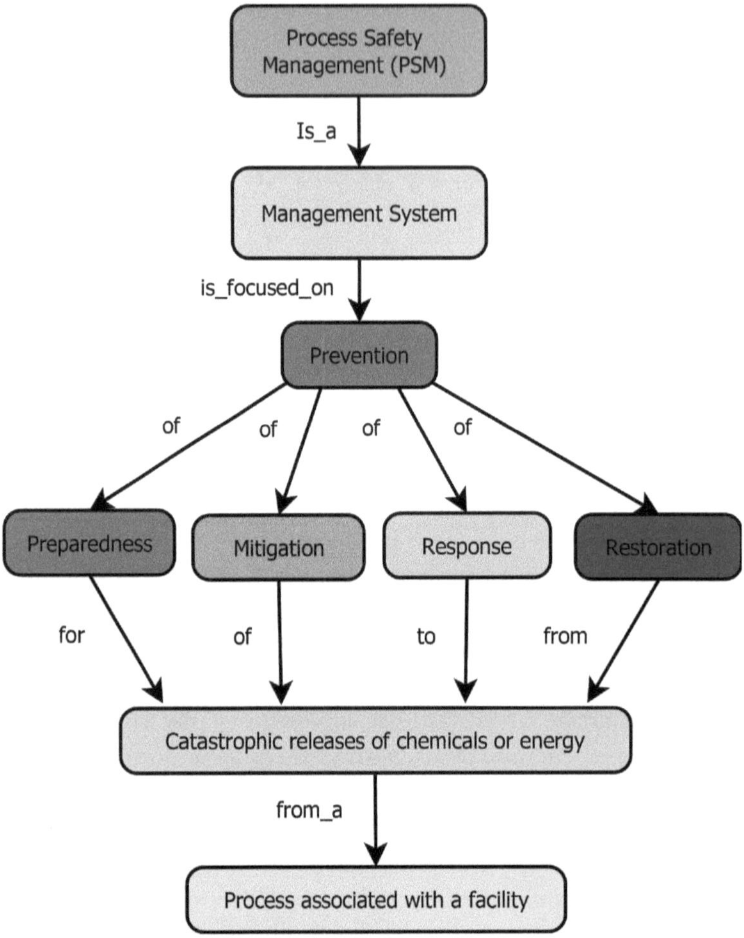

Fig. 7.1 Process safety management system.

the operating modes of the main subsystems, as well as the selection of optimal process control programs.

- The principle of hierarchical organization. Within the framework of this principle of designing control systems for large systems, it is assumed that the construction of a control system is carried out in the class of multi-level hierarchical control systems with division into control levels. Each level of the management system has its own management goals and methods for their achievement and implementation.

Control and management systems developed on the basis of these principles are called hierarchical control systems (Mesarouic, Macko, & Takahara, 1970).

Turning to the description of the architecture of the hierarchical control system, first of all, we note that this system, on the one hand, contains a number of subsystems, and on the other hand, it itself is part of a supersystem. Note that during system integration, the supersystem forms the goal of the subsystem.

To ensure systemic integrity of the control system, all subsystems of a hierarchically organized control system carry out an active interlevel exchange of information (see Fig. 7.2).

Fig. 7.2 uses the following notation:

$S_0 = US_i$ is the union of the hierarchical system and its external environment into one system;

f_i is environmental influences on various levels of the hierarchical system ($i = 1, 2, 3$); and

v_i is the impact of the hierarchical system on the external environment ($i = 1, 2, 3$).

The top level of the hierarchical management system is the organizational level or the planning level. The main tasks of this level include:

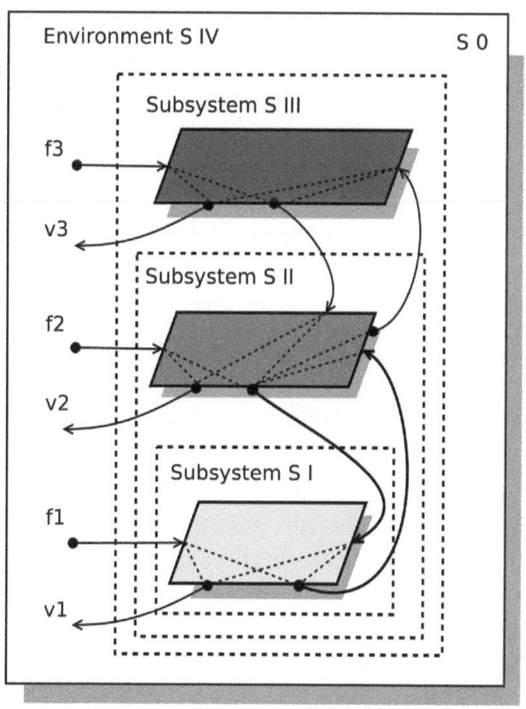

Fig. 7.2 Multilevel hierarchical control and management system.

- selection of a global management goal and its correction when the current management situation changes, including in case of emergency situations;
- optimization of the choice of control programs and ensuring information exchange with a data processing center;
- analysis and forecast of the general management situation in the conditions of a priori and a posteriori uncertainty; and
- formation of a system of management commands in the event of an emergency situation with the formation of response scenarios to the situation.

The following is the level of coordination that provides the solution to the following tasks:

- adaptation of the characteristics of the control system to changes in the external and internal situation, for example, the impact of external disturbing factors, equipment failures, etc.; and
- coordination of the work of subsystems of the lower level and the possible reconfiguration of these subsystems when the control situation changes.

The lower level of management system is the executive level, which provides:

- control of technological processes as a multiply connected nonlinear dynamic control object and its subsystems on a variety of standard and possible emergency conditions; and
- development of external disturbances: signal, parametric, structural.

Other basic principles in the construction of hierarchical control and management systems include:

- the principle of combining models, methods, and algorithms for the analysis and synthesis of multifunctional control systems, which consists of applying both classical methods of the theory of multivariable connected continuous and discrete automatic control systems, and intelligent control methods based on neural networks, fuzzy logic, genetic algorithms, etc.; and
- the principle of constructing open information management and control systems based on the intellectualization and standardization of information processing technologies at various stages of the life cycle of a large scale infrastructure system.

Note that to ensure high reliability and efficient functioning of hierarchical control and management systems at various stages of their life cycle, it is necessary to develop effective methods for managing the distributed computing resources of an enterprise.

7.2 Intelligent technologies for process safety

The tasks of analysis and synthesis of a risk management system in the class of hierarchical intelligent systems have a number of their own characteristics. Let's consider some of them.

Decomposition according to the hierarchy levels of the control system leads to a system of relatively independent control subsystems whose properties will change vertically in the hierarchical control system. In accordance with the IPDI principle proposed by D. Saridis, information flows from level to level differ in the content and pace of data generation (Saridis, 2001).

At the lower level, data obtained from various sensors and the results of operator actions are generated. At the coordination level, decisions are made to change the scenarios based on analyses of data obtained from the coordination level. An example of the executive level is a data acquisition and control system based on PLC (programmable logic controllers) and SCADA (supervisory control and data acquisition) systems in automatic mode. An example of the functioning of the coordination level is the joint work of its SCADA operators in dialogue mode.

At the planning level, new scenarios are formed, which are transferred to the coordination level. The formation of scenarios for solving production problems requires the processing of a large amount of data. In case of critical situations on the basis of preformed scenarios, operational actions are taken to minimize risks at the coordination level. An example of a planning level is the work of analysts, enterprise managers, and management personnel.

Communication between management levels in a hierarchical management system is determined by the rules and regulations of interaction. For technical subsystems, interaction protocols are built on the basis of various standards. For organizational systems, the rules for the exchange of information are determined by regulations and are based on workflow systems.

Big data technologies make it possible to form forecasts of the development of critical situations on the basis of processed data, based on which it is possible to develop risk minimization scenarios. The accuracy of forecasts can be quite high, since they are based on actual material, taking into account many production factors.

When solving the problems of designing risk minimization systems, one should take into account the mixed nature of the design decisions used at various levels of the hierarchical control system when building an integrated management system. Assessing the quality and effectiveness indicators of the hierarchical management system as a whole is a difficult task in itself, since it requires the analysis of a large amount of data. This is consistent with the

Ashby principle, since the complexity of the control system corresponds to the complexity of the control object (Ashby, 1956).

The specifics of the functioning of the hierarchical control system include the fulfillment of the requirement to ensure the stability of control processes, both under normal operating conditions and under the influence of uncertainty factors (for example, when subsystems fail or possible structural defects appear in distributed databases and knowledge databases).

Certain features of hierarchical control systems significantly complicate the task of designing and studying the risk management system of an industrial enterprise. Currently, hopes are pinned on artificial intelligence technology.

Artificial intelligence technologies are based on statistical training methods (Vapnik, 1998). Learning algorithms are divided into learning with and without a teacher. In the first case, a method is implemented based on the formation of a textbook with examples. Examples and answers from the textbook are loaded into the knowledge base, on the basis of which, using the learning algorithms, the classification system is set up. In the second case, the problem of classifying the presented examples is solved on the basis of algorithms that divide them into classes.

The formation of the textbook depends on the problem that is solved with the help of artificial intelligence technologies. But in any case, you need a sufficient number of examples to effectively solve the problems of approximation and classification (Theodoridis, 2020). Big data collected in the enterprise is an invaluable source of examples, situations, and scenarios. The task of creating a textbook is independent and is not considered in this chapter.

Computing methods that use neural networks, fuzzy logic, genetic algorithms, swarm intelligence, and similar methods for solving computational problems are called soft computing (Jiao, Shang, Liu, & Zhang, 2020; Leardi, 2020; Yang, 2020).

Let us further consider the main approaches used in hierarchical production risk management systems built on the basis of soft computing technologies.

In Fig. 7.3, a simplified structure of a hierarchical risk management system and methods for describing (models) management processes are presented, where EL (execution level), KL (coordination level), and PL (planning level) indicate the executive, coordination, and organizational management levels, respectively; FLS (fuzzy logic system), NFS (neuro-fuzzy system), ANN (artificial neural networks), DNN (deep

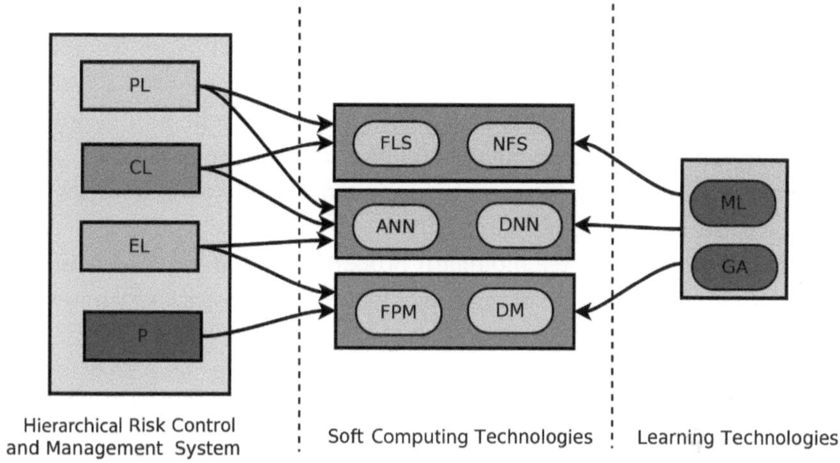

Fig. 7.3 Hierarchical control and management system and types of models of control and management processes.

learning neural networks), FPM (first principal models), and DM (digital model) designate models based on fuzzy logic, neural networks, and discrete and continuous dynamic models. As learning algorithms, algorithms based on machine learning (ML) methods and genetic algorithms (GA) can be used.

The considered class of systems belongs to the class of control systems of hybrid logical-dynamic systems (Han & Sanfelice, 2020).

Finding a solution for these systems is an extremely difficult task, since when designing control systems, it is necessary to take into account the nonlinear and discrete nature of control processes, the features of their interaction, the dynamics of the behavior of a multilevel system, etc.

7.2.1 Fuzzy risk control systems

Fuzzy control systems usually include a set of rules based on the experience of personnel and are part of the process regulations. Observed data from sensors and other sources of information are collected and used to evaluate these rules (Zadeh & Polak, 1969).

If the rules are logically satisfied, the pattern is identified, and the state of a process associated with that pattern is suggested. Each particular problem might imply a known control action.

Fuzzy systems allow us to make calculations with abstract values. To achieve this, the membership functions are applied. Each function can represent one abstract value, for example "small," "medium," or "large." All these functions provide the so-called fuzzification procedure, or conversion from numbers to abstract objects (see Fig. 7.4).

The advantage of this approach is the possibility of synthesis of algorithms of intellectual risk management as a complex dynamic object in the class of decision rules (products), based on the experience and knowledge of experts (Valeev & Kondratyeva, 2015).

When analyzing the technological process dynamics we can transform the position of technologic process state to the abstract space of parameters (see Fig. 7.5).

Here we have different trajectories of process dynamic marked by numbers from one to five:

- Trajectories 1-2-3 and 1-2-1 are normal processes.
- Trajectory 1-2-4 is a hazardous process.
- Trajectory 1-2-5 is a fault process.

When we analyze the space state of process (in Fig. 7.5) we can convert the results of our analysis to the set of rules (rules/knowledge base):

Rule 1: If (Pk = S) and (Ps = S) then (Process = Regular Area)
Rule 2: If (Pk = L) and (Ps = M) then (Process = Regular Area)
Rule 3: If (Pk = L) and (Ps = L) then (Process = Regular Area)
Rule 4: If (Pk = L) and (Ps = S) then (Process = Hazardous Area)

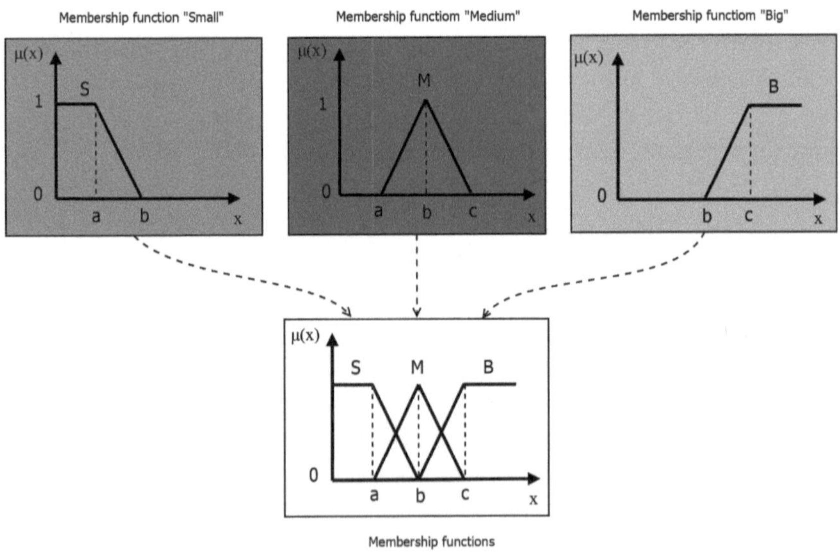

Fig. 7.4 Memberships functions.

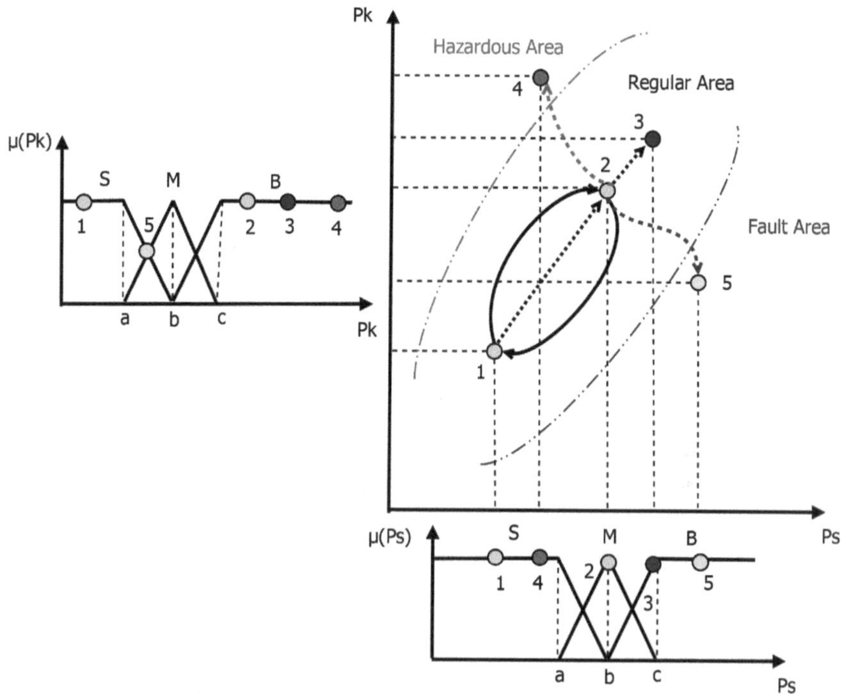

Fig. 7.5 Process space and fuzzy membership functions.

Rule 5: If (Pk = S or Pk = M) and (Ps = L) then (Process = Fault Area)
This is direct process of knowledge discovery. The same but inverse procedure of risk analysis with the knowledge base is presented below:

Rule 1: If (Process = Regular Area) then (Pk = S) and (Ps = S)
Rule 2: If (Process = Regular Area) then (Pk = L) and (Ps = M)
Rule 3: If (Process = Regular Area) then (Pk = L) and (Ps = L)
Rule 4: If (Process = Hazardous Area) then (Pk = L) and (Ps = S)
Rule 5: If (Process = Fault Area) then (Pk = S or Pk = M) and (Ps = L)

The result of transfer from process state space to fuzzy risk space is presented in Fig. 7.6, where RL is risk level space, H is hazard area, R is regular area, and F is failure area. The symbol "—" represents unknown area.

To cover risk level space RL we need nine rules, in the case of three membership functions. If we apply seven membership functions for more accurate calculations, we need 49 rules. In the case that our state space consists of three parameters and the fuzzy system includes seven membership functions for each parameter presentation, the rule base includes 343 rules.

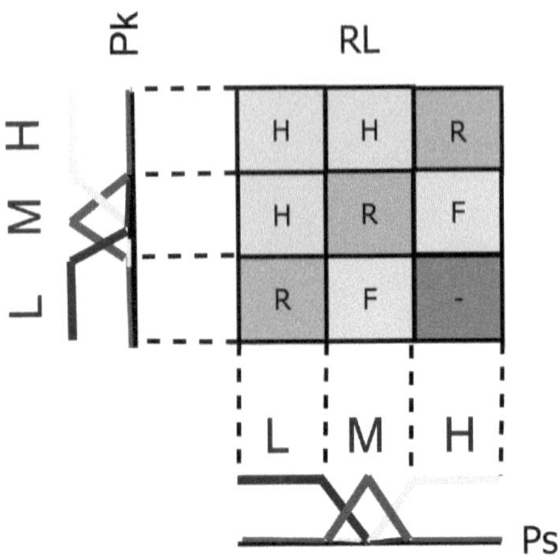

Fig. 7.6 Fuzzy risk space.

In some fuzzy systems, we need to find the appropriate form of membership functions. This procedure deals with learning process, or adaptation of fuzzy systems, to be more accurate.

If we need to create a system for risk assessment, we can apply various software tools. Fig. 7.7 shows the model of the fuzzy system that was

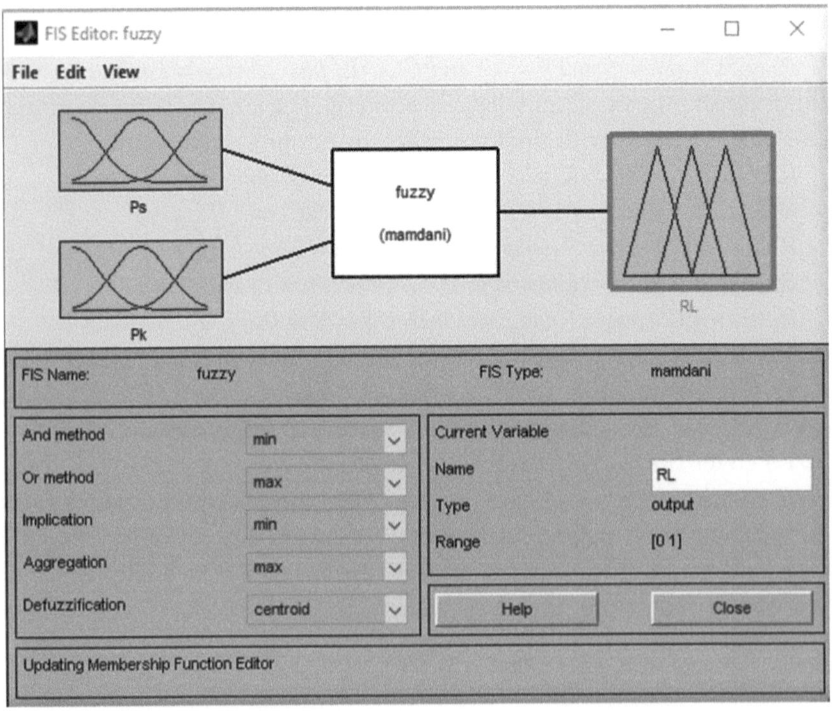

Fig. 7.7 Fuzzy inference system editor.

discussed earlier (see Fig. 7.5), created using the Matlab Fuzzy Toolbox application (MathWorks official website, n.d.). This software tool is useful for prototyping of fuzzy risk evaluation models and has a user-friendly interface.

To create a system for assessing fuzzy risk, we use the editor of the fuzzy inference system (see Fig. 7.7). Our system has two inputs, Ps and Pk, and an output, RL. As the core of the output, we use the Mamdani output method (Mamdani & Assilian, 1975).

To solve the fuzzification task (the transition from real values of input parameters to linguistic variables), three fuzzy membership functions are used, the form of which is shown in Fig. 7.8.

Note that the type of functions used in our system is chosen arbitrarily. Further, in the process of setting up a fuzzy system, the form of functions for each of the linguistic variables can change. Currently, various algorithms for selecting functions have been developed. The use of big data allows the use of large amounts of data to configure functions, which can improve the efficiency of a fuzzy risk analysis system.

Our fuzzy system include nine rules, two input variables with three membership functions and output with three membership functions (see Fig. 7.9).

The result of developing a fuzzy risk assessment system is the ability to build a nonlinear risk assessment function. For our example, this is a three-dimensional continuous function, the form of which is shown in Fig. 7.10.

7.2.2 Neural networks

Neural network technologies are widely used to solve various problems of approximation, pattern recognition, and classification (Snider & Yuen, 1998; Zadeh, 1994). Neural networks belong to the class of universal approximators (Winter & Widrow, 1988). This means that you can always find the structure of the neural network, which will solve the problem of approximating a given function with the required accuracy. Function values are presented in tabular form.

If we represent a neural network in the form of a "black box" model, then the neural network Fnn(•), when solving the training problem (setting the weights of synaptic connections):

$$Ynn = Fnn\,(X),$$

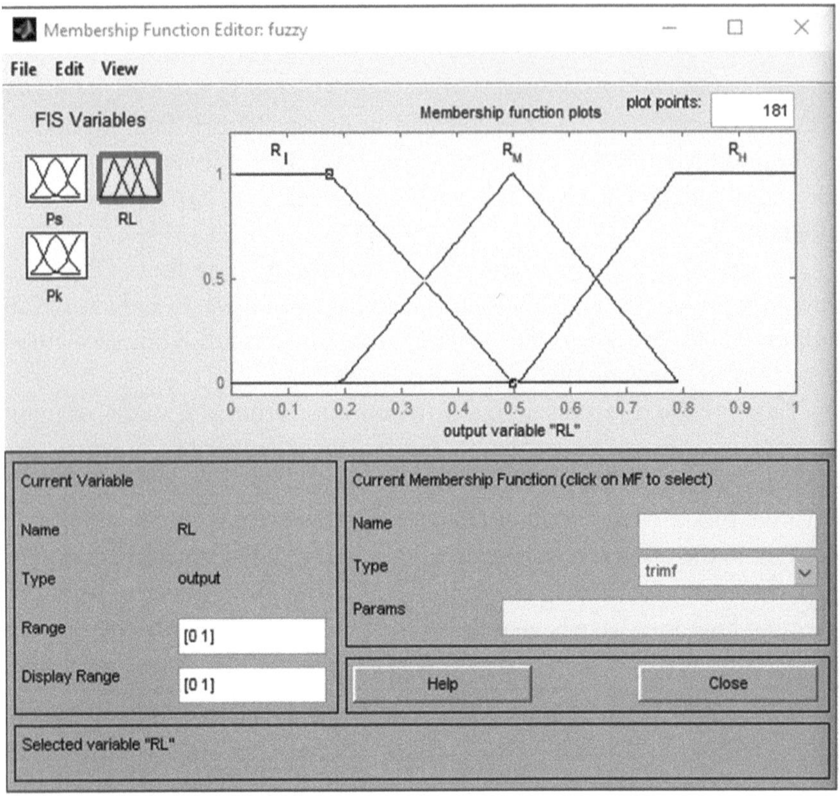

Fig. 7.8 Membership function editor.

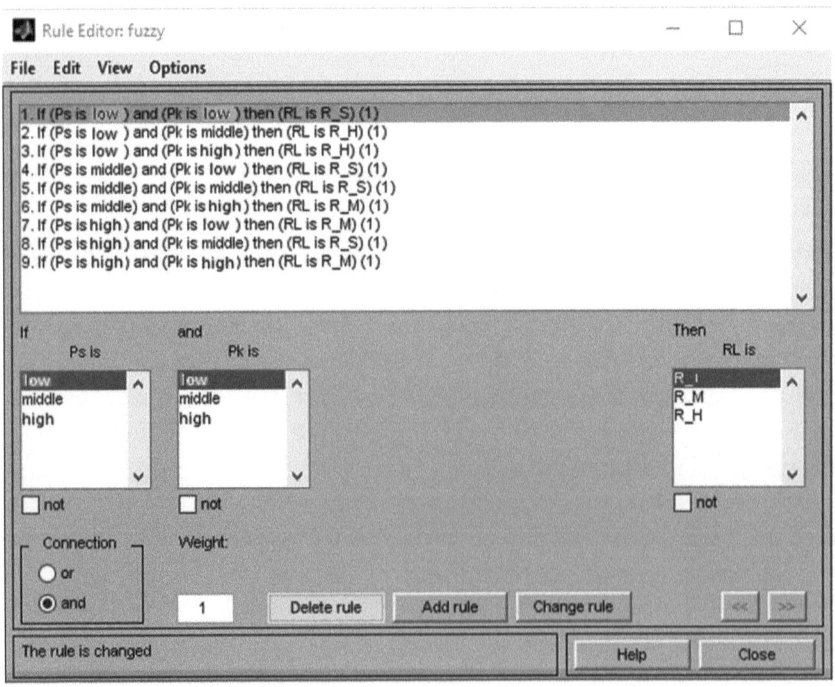

Fig. 7.9 Rule editor.

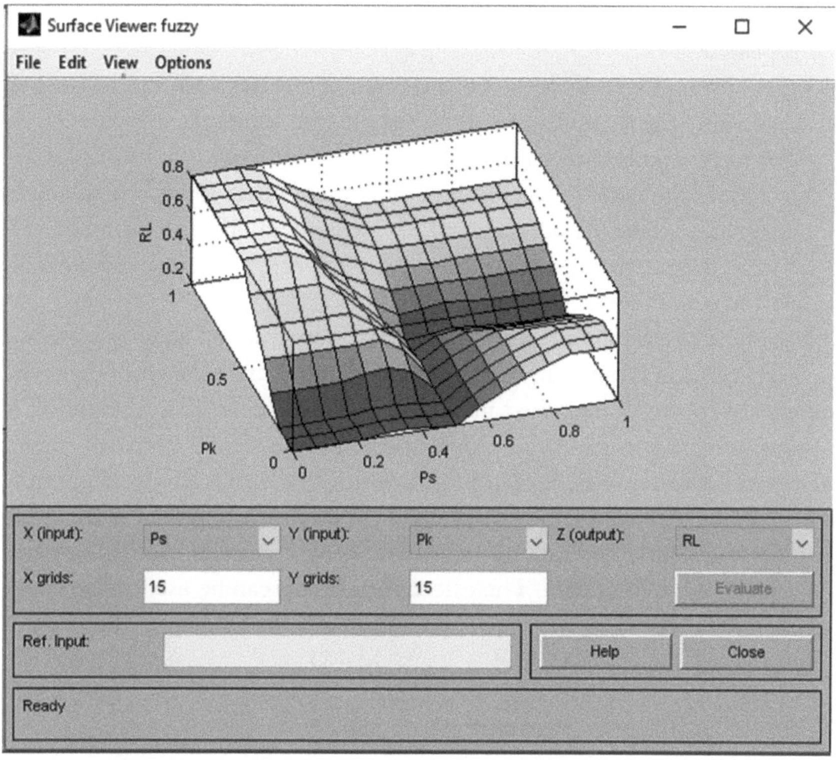

Fig. 7.10 Fuzzy surface viewer.

where X is the vector of input variables from the textbook and Ynn is the vector of output values obtained using a neural network.

The following condition must be met:

$$\min\left(\text{Ynn} - \text{Y}\right),$$

where Y are the real values used in training the neural network.

A neural network has its own life cycle:

- Design phase. At this stage, the expert selects the type of neural network, for example, a multilayer perceptron. An array called a textbook is formed from the available data, including an array of input-output pairs. The dimension of the input and output vectors of the textbook may vary. Using the training algorithms, the neural network coefficients F(•) are tuned. If the accuracy of the neural network approximation satisfies the specified accuracy, then the trained neural network can be used during the operation phase, otherwise all procedures will need to be repeated.

- Operation phase. At this stage, a trained neural network can be used to determine the desired value, for example, the value of the risk level. At this stage, taking into account the formation of new data, it is necessary to retrain the neural network, i.e., the design phase is performed anew.
- Adaptation stage. At this stage, taking into account the formation of new data, it is necessary to retrain the neural network, i.e., the design phase is performed anew.

Fig. 7.11 shows the architecture of a neural network risk analysis system.

The vector of input signals X and the vector of output values of the technological process Y are stored in databases. From these data, a textbook is formed, which is used in the process of training a neural network. Based on expert estimates of R, a vector RL is formed, which is also used in training the neural network and generating the vector of weight values W. After training, the neural network can calculate the risk value R_1 in real time on the basis of X_1 and Y_1 data. The calculation results can be used at the coordination level to select the appropriate scenario for managing the current situation.

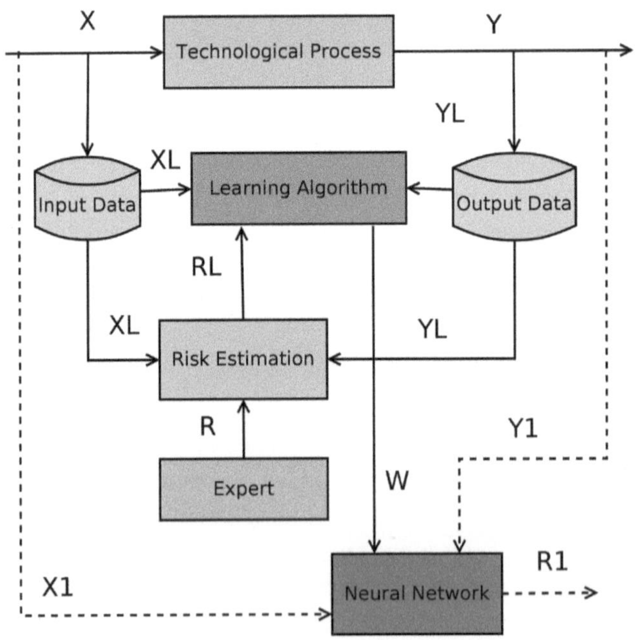

Fig. 7.11 Architecture of a neural network risk analysis system.

7.2.3 Expert systems

For using experts' knowledge in the field of process safety, when risk control system is designed, expert system technologies can be applied. An expert system is a set of rules that are used to describe certain process state patterns (Leondes, 2002).

The architecture of an expert management system is shown in Fig. 7.12. The system under consideration consists of two hierarchical levels: the level of data collection and generation of command signals, and the level of decision support based on expert knowledge. Knowledge can be represented as an ordered set of logical rules.

The difference between the system under consideration and the three-level control system is that the system is focused on solving a specific problem for which it is possible to form a set of rules based on expert knowledge or regulations.

Further, we consider the features of the design procedure of risk management algorithms for the system under consideration. Based on the knowledge of the developers of the control system, a knowledge base is formed in which the logical rules for choosing control algorithms for various control situations are stored. When performing this stage of system development, problems arise associated with the formation of the information-logical structure of the knowledge base and the architecture of the inference machine.

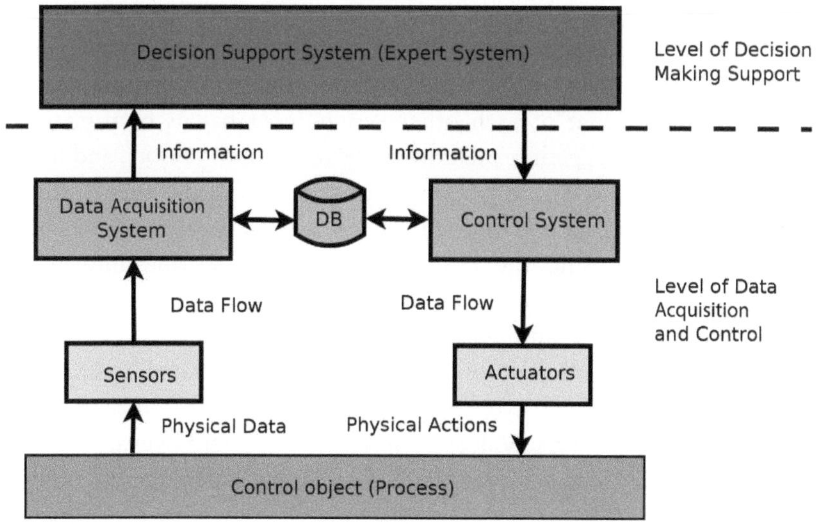

Fig. 7.12 Decision support system architecture.

It should be noted that it is necessary to use different information models for each of the possible critical situations to identify the general patterns of distribution of information resources and information flows.

The knowledge base of the decision-making support system under consideration should contain rules that allow the transition from the space of management tasks to the space of risk minimization scenarios. Rules can be formed based not only on expert knowledge, but also on the results of evolutionary design procedures using big data technology.

Of great importance in the implementation of such systems is the optimization of the volume of the knowledge base and its structural verification, since the number of rules in the knowledge base depends on the a priori available amount of information on process safety and a set of risk minimization algorithms.

7.2.4 Multiagent systems

Active research is currently underway in the field of building distributed control systems based on the multiagent paradigm (Zhu et al., 2020). The basis of this paradigm is the idea of building a management and decision-making system as a distributed community of "agents" who independently make decisions and, if necessary, can use the "collective mind" to achieve their goals under conditions of significant uncertainty.

The generalized architecture of a multiagent system is shown in Fig. 7.13. A multiagent system includes two active agents that have the ability to analyze the state of the information field of a distributed control object.

A feature of such an organization of the management system is that the agents are independent of each other—each agent has its own goal. To achieve this goal, the agent uses information from input sensors, and if necessary, can use actuators. To perform targeted actions, the agent uses the built-in knowledge base and logical inference system. Agents, in the course of their activities, using the communication channel and information exchange protocols, can exchange data, knowledge, and control signals. In real applications, the number of agents is not limited.

One area of application of multiagent systems is associated with the creation and use of intelligent systems for collecting information using intelligent sensors and actuators (Valeev, Taimurzin, & Kondratyeva, 2013). In this case, intelligent sensors are understood as multifunctional programmable

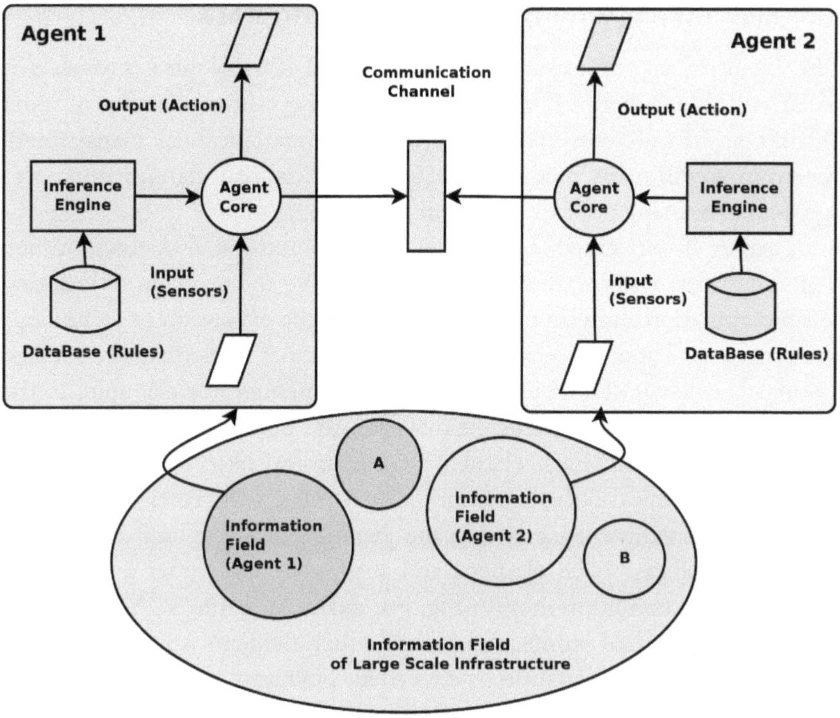

Fig. 7.13 Multiagent system.

measuring instruments equipped with microprocessors, having the ability to connect to a data network and, if necessary, change their location. Devices designed to form control actions and endowed with certain intellectual abilities are called smart actuators.

The set of functions of smart sensors and smart actuators includes functions such as conversion of primary information, self-monitoring and self-diagnosis, signal preprocessing, reconfiguration in case of emergency, etc.

A feature of multiagent systems is the cooperation of agents when it is necessary to achieve their goals. If the agent cannot obtain reliable information about the state of the infrastructure object in any part of it, and this information is important for the agent to achieve their goal, they can use the resources of another agent. For example, in Fig. 7.13, these are the tasks of collecting information in areas A and B. Agents can use their actuators to change their position and can transmit additional information received to each other via a communication channel.

7.3 Risk management systems and big data

The hierarchical risk management system for technological processes is based on the paradigm of the data ecosystem. Let us consider this in more detail. Fig. 7.14 presents the architecture of a hierarchical risk management system in which many processes of data collection, data transmission, storage, and processing are carried out in real time.

Chapter 1 examined various technical objects and systems. When designing these systems, developers have to solve the problem of ensuring reliable operation and minimizing risks during the execution of technological processes. These objects include a large number of elements connected among themselves. These joints can be implemented, for example, in the form of welds or bolted joints. In this case, the objects are combined into technical subsystems. Each element of a technical object is accompanied by a set of technical documentation, which reflects the rules of operation of the object, maintenance within the product life cycle, and repair rules in case of failures. Each element and subsystem undergoes a scheduled inspection and condition monitoring. For example, in the SCADA system, the operating time of pumps, valves, and other elements is specified. If the operational life of the facilities is exceeded, operators are given a message about the need to repair or replace equipment. These procedures are accompanied by a set of documents that are stored in paper form and in the form of data in the SCADA system database. When analyzing the state of various structures, it is necessary to monitor the state of welds and fasteners, the condition of the cables, etc. The results of the planned inspection are entered in paper forms and, in the form of data, are transferred to the technical system maintenance databases. These data can be used to assess the state of the elements and form a forecast of their performance.

In Chapter 1, technical systems were presented in the form of a graph that included elements and the relationships between them. The destruction of elements and relationships leads to the destruction of the system. Thus, in one case or another, various risks arise that lead to a critical situation. For example, the breakage of an oil rig column cable can lead to deaths and destruction of drilling equipment. Another example is the violation of the quality of the welds of the base of an offshore drilling platform, which, in turn, causes the development of a critical situation. Prediction of such situations on the basis of available data collected in the data warehouse allows you to build a predictive model and, in some cases, prevent the occurrence of a critical situation. Fig. 7.14 schematically presents the procedures for

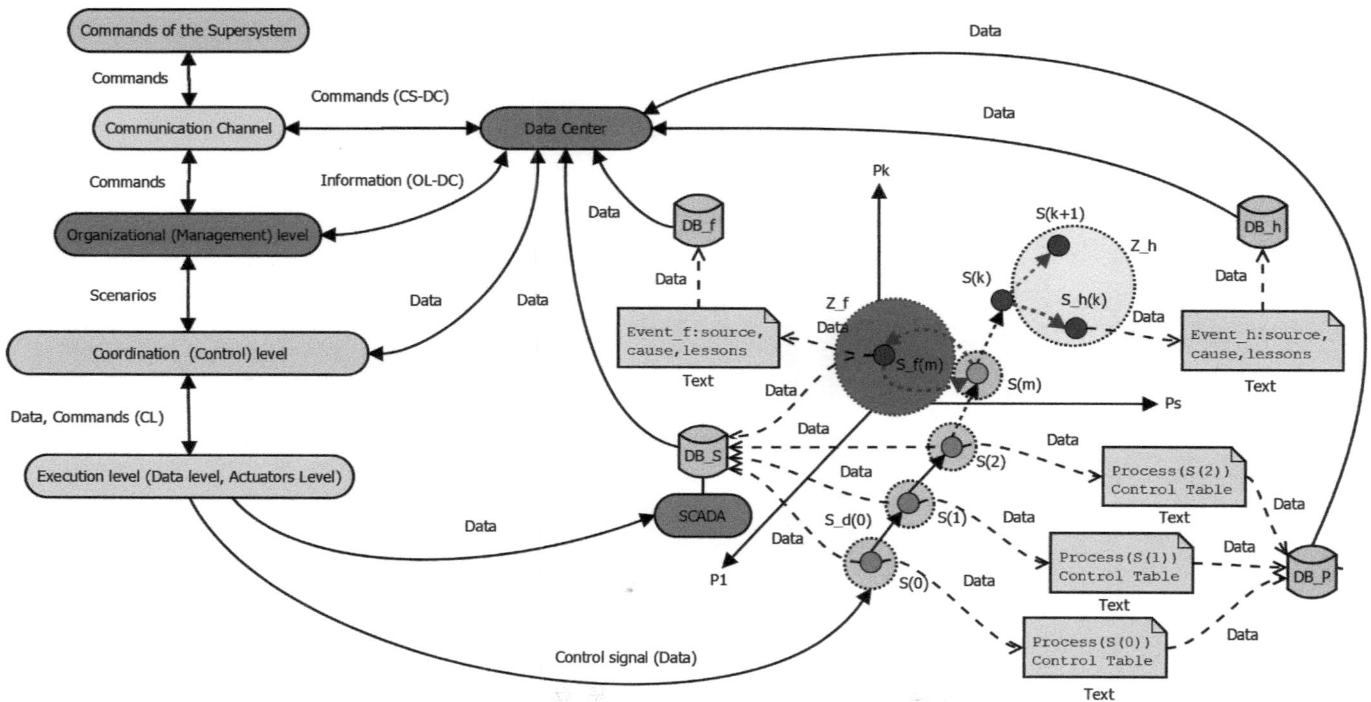

Fig. 7.14 Hierarchical safety management system.

generating arrays of data on the state of technological processes and technological equipment, which are stored in the DB_P database. Data on failures, their causes, and measures taken are stored in the DB_f database. Data on the current state of technological processes and operator actions is stored in the SCADA database of the DB_S system. The data obtained during the analysis of critical situations, the history of events, and the conclusions of experts is stored in the database DB_h.

Data from these databases is transferred to the data warehouse. Modern infrastructure facilities operate under the influence of various uncertainty factors, both internal and external. Internal factors of uncertainty include depreciation of equipment, violation of work regulations, personnel errors, and equipment failures. External factors of uncertainty include the environmental impact on equipment and the health status of personnel.

All these factors influence technological processes and can cause a critical situation.

To analyze and control the state of the infrastructure object, an internal audit and an external audit are performed. In an internal audit, based on the available data, the auditor, based on the internal regulations of the enterprise, determines the current state of the enterprise on the topic of audit. To solve these problems, all available information is used, presented in the form of sets of documents and statistical data. Usually, this data is stored on the company's servers.

During an external audit of the company, compliance with standards on the audit topic is checked. Standards in the field of process safety were discussed in Chapter 2. The standards contain recommendations, rules, and methods for achieving the goal indicated in the standard. To solve the problems of external audit, a large amount of information is required, which is stored on the company's servers and in the company's reports made publicly available. Compliance with the standards of those or parties of the modern enterprise has a positive effect on the image of the enterprise and decisions of insurance companies.

In particular, when conducting a PSM audit, the auditor should have a wide range of data. First, information is required on the hazards of chemicals used in the process. During the audit, the most accurate and complete information in writing (handwritten or electronic) about chemicals, technological equipment, and technological processes is used. The information collected on chemicals makes it possible to evaluate the characteristics of possible fires and explosions, the dangers of reactivity, and the effects of corrosion and erosion on technological equipment and monitoring tools, as well as the health risks of workers.

Typically, this is based on the data of the material safety data sheet (MSDS), as well as data on the chemical composition of the process, including the reaction to escape and the danger of overpressure, if required at the plant. Process information should include criteria established by the employer for the maximum stockpile of process chemicals, the limits beyond which the violated conditions will be considered, and a qualitative assessment of the consequences or results of deviations that may arise when working outside established technological constraints. In addition, information is needed about the entire chain of equipment that provides a particular chemical process.

Various tools are used to visualize technological processes. The flow block diagram is a simplified diagram and is used to display the main processing equipment and interconnected production lines and flow rates, flow composition, temperatures, and pressures. Flow charts are more complex and show all the main flows. In this case, the inlet and outlet of collectors and heat exchangers, as well as valves, are also taken into account, which facilitates understanding of process control. In addition, the corresponding pressure and temperature measurement points on all supply and product lines in all major containers are indicated. Process flow charts typically show the main components of control loops along with key components of the production chain. Piping and instrument diagrams are used for a more detailed analysis of the relationships between equipment and instrumentation.

The information necessary to solve management problems in the field of process safety is generated from various data sources and aggregated information from various information systems. Chapter 3 examined various sources of data on the state of process parameters. The data obtained is used in solving management problems and monitoring the status of equipment and process parameters. Based on these data, a conclusion is made about the level of risks and their change.

Enterprises generate a large amount of data in the databases discussed in Chapter 4. Features of data arrays determine the type of databases that are used to store this data. It should be noted that the data is generated in real time and can be stored for many years. There are problems maintaining archives in working condition. One of the ways is the use of cloud storage technologies implemented by the enterprise or a service provider.

When processing large amounts of data, the time to achieve the desired result is increased. Large data processing technologies are based on big data software ecosystems and the use of high-performance cluster systems. Note that the maintenance of software and server equipment requires highly qualified developers, programmers, and maintenance personnel.

Errors in the development and maintenance of these systems can cause critical situations. Given the complexity of the equipment used and the processes themselves, data collection on all possible conditions is not always possible.

Thus, the data required for the PSM audit provides the foundation for process safety management. The data are presented in various forms, including handwritten and electronic documents, drawings, questionnaires, and employee testing results in the form of special files, standard forms with the results of incident investigations, image files, as well as audio and video recordings.

Chapter 5 discussed the features of digital counterparts with the help of which digital models of technological equipment are formed.

Large volumes of data stored in databases are a digital copy of the processes associated with the current situation at the enterprise, and also store the state in the state space of technological processes, including emergency situations and failure situations. To process this information, analytical systems are used, with the help of which forecasts are made for changes in various indicators of the enterprise.

Modeling can also serve as a generator of missing evidence for process safety risk management. This is especially true when process industry enterprises are not able to provide the collection of large sets of real data. Adequate simulation models of real production facilities can be considered as virtual enterprises that simulate data generation in formats identical to real sensors in real production.

Chapter 6 discussed the basic technologies, methods, and algorithms used in analytical systems. Big data technologies make it possible to extract knowledge from data arrays and increase the accuracy of risk assessment procedures through training methods.

7.4 Summary

The problem of risk management of infrastructure facilities is associated with the need to solve monitoring and control tasks in real time. The emergence of critical situations also requires real-time decision making. Unfortunately, it is not always possible to avoid the effects of harm to human health and life, as well as great damage and harm to the ecological state of the environment. In some cases, it is possible to predict the emergence of causes that can cause a critical situation. To solve this problem, it is necessary to process a large amount of data and apply knowledge management systems.

The main artificial intelligence technologies that can be effectively used to assess current risks are considered. As an example, the use of fuzzy logic in designing risk assessment systems for identifying various states of the technological process: normal functioning, the occurrence of failures and the critical situations, is considered.

In decision-making systems, expert systems are traditionally used. These systems use the knowledge of experts presented in the form of production rules. An expert system allows you to get a solution and show the procedure for finding an answer.

The distributed nature of the infrastructure facility imposes features on the organization of management processes. The chapter discusses the technology of multiagent systems, which allow the efficiency of the development of the control system and its maintenance to be increased. At the end of the chapter, the architecture features of the hierarchical system of control and risk management of technological processes are discussed.

7.5 Definitions

Risk control or hazard control is a part of the risk management process in which methods for reduction of identified risks are implemented.

Neural networks are computing systems that "learn" to perform tasks by considering examples. Fuzzy logic is a form of many-valued logic in which the truth values of variables may be any real number between 0 and 1.

Expert system is a computer system that emulates the decision-making ability of an expert.

Multiagent system is a system composed of multiple interacting intelligent agents.

Decision support system is an information system that supports business or organizational decision-making activities.

References

Ashby, W. R. (1956). *An introduction to cybernetics*. Chapman & Hall.

Han, H., & Sanfelice, R. G. (2020). Linear temporal logic for hybrid dynamical systems: Characterizations and sufficient conditions. *Nonlinear Analysis: Hybrid Systems, 36*. https://doi.org/10.1016/j.nahs.2020.100865, 100865.

Jiao, L., Shang, R., Liu, F., & Zhang, W. (Eds.). (2020). *Brain and nature-inspired learning computation and recognition* Elsevier. https://doi.org/10.1016/B978-0-12-819795-0.01001-X.

Leardi, R. (2020). Genetic algorithms in chemistry. In *Comprehensive chemometrics* (2nd ed., pp. 617–634). Elsevier. https://doi.org/10.1016/B978-0-12-409547-2.14867-X.
Leondes, C. T. (Ed.). (2002). *Expert Systems, Vol. 1–6.* Academic Press. https://doi.org/10.1016/B978-012443880-4/50045-4.
Mamdani, E. H., & Assilian, S. (1975). An experiment in linguistic synthesis with a fuzzy logic controller. *International Journal of Man-Machine Studies*, 7(1), 1–13.
MathWorks official website (n.d.). Retrieved 18 April 2020, from: https://www.mathworks.com/.
Mesarouic, M. D., Macko, D., & Takahara, Y. (1970). *Theory of hierarchical, multilevel, systems* (1st ed.). Elsevier Science.
Saridis, G. N. (2001). *Hierarchically intelligent machines.*
Snider, L. A., & Yuen, Y. S. (1998). The artificial neural-networks-based relay algorithm for the detection of stochastic high impedance faults. *Neurocomputing*, *23*(1), 243–254. https://doi.org/10.1016/S0925-2312(98)00068-X.
Theodoridis, S. (2020). *Machine learning* (2nd ed.). Academic Press. https://doi.org/10.1016/B978-0-12-818803-3.00004-0.
Valeev, S., & Kondratyeva, N. (2015, August). Technical safety system with self-organizing sensor system and fuzzy decision support system. In *Presented at the 2015 IEEE international conference on fuzzy systems (FUZZ-IEEE).* Istanbul, Turkey: IEEE. https://doi.org/10.1109/FUZZ-IEEE.2015.7337962.
Valeev, S. S., Taimurzin, M. I., & Kondratyeva, N. V. (2013). An adaptive data acquisition system in technical safety systems. *Automation and Remote Control*, *74*, 2137–2142. https://doi.org/10.1134/S0005117913120151.
Vapnik, V. N. (1998). *Statistical learning theory.* NY: John Wiley.
Winter, R., & Widrow, B. (1988). MADALINE rule II: A training algorithm for neural networks. In *1. INNS first annual meeting* (p. 148). https://doi.org/10.1016/0893-6080(88)90187-6.
Yang, X.-S. (Ed.). (2020). Nature-inspired computation and swarm intelligence. In *Algorithms, theory and applications* Academic Press. https://doi.org/10.1016/C2019-0-00628-0.
Zadeh, L. A. (1994). Fuzzy logic, neural networks, and soft computing. *Communications of the ACM*, *37*(3), 77–84.
Zadeh, L. A., & Polak, E. (1969). *Toward a theory of fuzzy systems.* New York: McGraw-Hill Book Co.
Zhu, Q. M., Wang, J., Wang, C., Xin, M., Ding, Z., & Shan, J. (Eds.). (2020). *Emerging methodologies and applications in modelling, cooperative control of multi-agent systems* Academic Press. https://doi.org/10.1016/B978-0-12-820118-3.00010-7.

Index

Note: Page numbers followed by *f* indicate figures, *t* indicate tables, and *b* indicate boxes.

E

F

FSC
www.fsc.org
MIX
Papier | Fördert gute Waldnutzung
FSC® C083411

Kolibri 360 GmbH
Ferdinand-Jühlke-Straße 7
99095 Erfurt, Deutschland
produktsicherheit@kolibri360.de